# Functional Biomaterials for Regenerative Dentistry

# Functional Biomaterials for Regenerative Dentistry

Guest Editor

**Nicholas G. Fischer**

MDPI

Basel • Beijing • Wuhan • Barcelona • Belgrade • Novi Sad • Cluj • Manchester

*Guest Editor*
Nicholas G. Fischer
Minnesota Dental Research
Center for Biomaterials
and Biomechanics
University of Minnesota
Minneapolis, MN
USA

*Editorial Office*
MDPI AG
Grosspeteranlage 5
4052 Basel, Switzerland

This is a reprint of the Special Issue, published open access by the journal *Journal of Functional Biomaterials* (ISSN 2079-4983), freely accessible at: https://www.mdpi.com/journal/jfb/special_issues/9HM3T339IB.

For citation purposes, cite each article independently as indicated on the article page online and as indicated below:

| |
|---|
| Lastname, A.A.; Lastname, B.B. Article Title. *Journal Name* **Year**, *Volume Number*, Page Range. |

**ISBN 978-3-7258-5279-6 (Hbk)**
**ISBN 978-3-7258-5280-2 (PDF)**
**https://doi.org/10.3390/books978-3-7258-5280-2**

# Contents

# About the Editor

**Nicholas G. Fischer**

Nicholas G. Fischer is a postdoctoral researcher at the University of Pennsylvania. His work has been recognized with funding from NIH/NIDCR through the F30 Kirschstein-NRSA program and he has received awards from the Institute for Engineering in Medicine, American Dental Association, and International Association for Dental Research, among others. Nick has published over 75 papers and serves on editorial boards for 11 journals. He holds a BSEvs from Creighton University and both a PhD and DDS from University of Minnesota.

Journal of
*Functional Biomaterials*

MDPI

*Editorial*

# Functional Biomaterials for Regenerative Dentistry

**Nicholas G. Fischer**

MDRCBB-Minnesota Dental Research Center for Biomaterials and Biomechanics, University of Minnesota, 515 Delaware St. SE, Minneapolis, MN 55455, USA; fisc0456@umn.edu

## 1. Introduction

Regenerative dentistry hopes to restore oral health by replacing diseased or damaged tissues with biologically functional, integrated counterparts. This differs critically from past approaches that rely on bonded and inert synthetic materials. The oral cavity contains a complex array of tissues frequently affected by disease, including the periodontal ligament, the dental pulp, the alveolar bone, and the mineralized tooth itself. The regeneration of each of these components presents significant scientific and clinical hurdles. Addressing these challenges is crucial for achieving oral health and improving patients' quality of life.

In this evolving landscape, biomaterials that stimulate and support tissue regeneration are slowly becoming indispensable tools in everyday dental practice. These materials' potential is derived from their ability to be tailored to individual patients, their compatibility with living tissues, and their capacity to modulate immune responses—qualities that position regenerative biomaterials as the cornerstone of future dental therapies. Here, we highlight work from a Special Issue showcasing cutting-edge research on functional biomaterials for regenerative dentistry.

## 2. Special Issue Highlights

Medication-related osteonecrosis of the jaw (MRONJ) is a destructive bone condition caused by certain drugs, such as antiresorptive and antiangiogenic drugs, in the absence of radiation treatment. Minimal definitive treatment options exist. Advanced platelet-rich fibrin (A-PRF) is a platelet-rich blood preparation with high concentrations of growth factors that has been suggested as a possible treatment option for MRONJ. Adamska et al. [1] retrospectively studied clinical data from 28 patients presenting with osteomyelitis due to MRONJ. The authors determined that less advanced lesions, in the absence of risk factors, had a better prognosis than advanced lesions and that A-PRF showed no significant effect on MRONJ resolution.

A-PRF has been evaluated in a variety of ever-expanding oral maxillofacial surgery applications. As a result, Chmielewski et al. [2] performed a meta-analysis focused on the question, "Does A-PRF provide better clinical outcomes than other materials used in exact oral and maxillo-facial procedures?" The results generally showed some benefit in a variety of clinical procedures, such as bone healing and postoperative pain reduction, albeit with limited evidence. Further analysis by Chmielewski et al. [3] focused on advanced platelet-rich fibrin + (A-PRF+). In comparison to A-PRF, A-PRF+ is produced with a different centrifugation protocol and is supposed to have a higher concentration of growth factors and a different fibrin network structure. The authors' systematic review showed that the performance of A-PRF+ was not significantly improved compared to other blood preparations, such as A-PRF, but there were slight trends toward improvement.

Decades of research have intensively investigated bone grafting materials. Dentin has been suggested as an alternative material given it may be autologous and is compositionally similar to bone. However, many dentin processing and preparation parameters remain poorly optimized. Mazzucchi et al. [4] examined the influence of tooth age on the dentinal release of growth factors important to bone growth. The authors showed that age did not affect growth factor release, albeit under the specific preparation parameters the authors used.

Clinicians frequently chose materials based on personal preferences and experiences. Friedmann et al. [5] compared three dental restorative centers that used either cross-linked high-molecular-weight hyaluronic acid (xHyA) with a xenograft, or enamel matrix derivative (EMD) with an allograft, or xHyA with a resorbable membrane for restoration of intrabody defects. The authors showed that, depending on the outcome variables, results were similar between groups, lending support to the use of xHyA compared to the classic EMD approach. Similarly, Panda et al. [6] compared plasma rich in growth factors (PRGFs) versus xenogenic bone graft (BXG) in patients for the regeneration of intrabony defects and showed similar results. The authors wisely noted "treatment decisions [should be] guided by patient-specific factors and clinical goals."

Park et al. [7] explored the idea of an oral probiotic, *Weissella cibaria*, for potentially protecting against tooth-supporting bone destruction from periodontal diseases. Periodontal diseases are common and marked, as they advance, by osteoclast-mediated bone destruction. The authors showed in vitro that *Weissella cibaria* suppressed osteoclast differentiation and activity. Further research on delivery methods may enable further deployment of probiotics in regenerative dentistry.

Dental caries are one of the most widespread infectious diseases in the world. Consistent evidence for decades has supported the widespread implementation of fluoride. Continual refinement of fluoride-containing dental materials has been motivated by patient preferences. Iwawaki et al. [8] evaluated a high-concentration fluoride varnish in comparison to a typical fluoride mouthwash in an early enamel caries in vitro model and showed the high-concentration fluoride varnish did not increase remineralization of the subsurface demineralized layer, but the varnish did improve the acid resistance of the model carious lesion.

The rising popularity of ceramic veneers as indirect esthetic restorations has fueled intense interest in the development of longer-lasting restorative materials. In particular, the wear resistance of dental materials in the highly biologically and mechanically active oral cavity has been historically evaluated to ensure long-term, esthetic results. Oshika et al. [9] showed the wear resistance of light-curable resin luting cements was similar to dual-cure cements and flowable resin-based composites.

Dentistry has been at the forefront of nanotechnology with the advances in filler technologies for dental resin composites. Nanoparticles have been explored for a variety of applications, especially those nanoparticles composed of gold. Jongrungsomran et al. [10] reviewed the use of gold nanoparticles in dentistry and concluded there was substantial promise for their use, particularly for "enhancing biological, mechanical and optical properties."

## 3. Concluding Remarks

By bringing together innovations in material science, biology, and clinical practice, regenerative dentistry is poised to redefine the standard of dental, oral, and craniofacial care. The contributions in this Special Issue advance the scientific understanding of functional biomaterials and accelerate their translation into routine dental procedures. As the boundary between restoration and regeneration continues to blur, the future of dentistry will be shaped by therapies that are not only reparative but truly regenerative—restoring both form and function in a patient-centered manner.

**Funding:** This manuscript received no external funding.

**Conflicts of Interest:** The authors declare no conflicts of interest.

## References

1. Adamska, P.; Stasiak, M.; Kobusińska, N.; Bartmański, M.; Zedler, A.; Studniarek, M. Treatment of Medication-Related Osteonecrosis of the Jaw Without and With the Use of Advanced Platelet-Rich Fibrin: A Retrospective Clinical Study. *J. Funct. Biomater.* **2025**, *16*, 180. [CrossRef] [PubMed]
2. Chmielewski, M.; Pilloni, A.; Adamska, P. Application of Advanced Platelet-Rich Fibrin in Oral and Maxillo-Facial Surgery: A Systematic Review. *J. Funct. Biomater.* **2024**, *15*, 377. [CrossRef] [PubMed]
3. Chmielewski, M.; Pilloni, A.; Adamska, P. Advanced Platelet-Rich Fibrin Plus (A-PRF+) as an Additive to Hard Tissue Managing Protocols in Oral Surgery: A Systematic Review. *J. Funct. Biomater.* **2025**, *16*, 145. [CrossRef] [PubMed]
4. Mazzucchi, G.; Mariano, A.; Serafini, G.; Lamazza, L.; Scotto d'Abusco, A.; De Biase, A.; Lollobrigida, M. Osteoinductive Properties of Autologous Dentin: An Ex Vivo Study on Extracted Teeth. *J. Funct. Biomater.* **2024**, *15*, 162. [CrossRef]
5. Friedmann, A.; Liedloff, P.; Eliezer, M.; Brincat, A.; Ostermann, T.; Diehl, D. Reconstructive Approach in Residual Periodontal Pockets with Biofunctionalized Heterografts—A Retrospective Comparison of 12-Month Data from Three Centers. *J. Funct. Biomater.* **2024**, *15*, 39. [CrossRef] [PubMed]
6. Panda, S.; Panda, S.; Das, A.C.; Lewkowicz, N.; Lapinska, B.; Tumedei, M.; Goker, F.; Cenzato, N.; Del Fabbro, M. Plasma Rich in Growth Factors Compared to Xenogenic Bone Graft in Treatment of Periodontal Intra-Osseous Defects—A Prospective, Comparative Clinical Study. *J. Funct. Biomater.* **2024**, *15*, 336. [CrossRef]
7. Park, G.-Y.; Park, J.-A.; Kang, M.-S. In Vitro Effects of Weissella Cibaria CMU and CMS1 on Receptor Activator of NF-κB Ligand (RANKL)-Induced Osteoclast Differentiation. *J. Funct. Biomater.* **2024**, *15*, 65. [CrossRef] [PubMed]
8. Iwawaki, R.; Horie, T.; Alhotan, A.; Nagatsuka, Y.; Sakuma, K.; Yoshihara, K.; Tsujimoto, A. Effect of Fluoride Varnishes on Demineralization and Acid Resistance in Subsurface Demineralized Lesion Models. *J. Funct. Biomater.* **2024**, *15*, 380. [CrossRef] [PubMed]
9. Oshika, M.; Kishimoto, T.; Horie, T.; Alhotan, A.; Irie, M.; Sule, V.C.; Barkmeier, W.W.; Tsujimoto, A. Wear Resistance of Light-Cure Resin Luting Cements for Ceramic Veneers. *J. Funct. Biomater.* **2024**, *16*, 5. [CrossRef] [PubMed]
10. Jongrungsomran, S.; Pissuwan, D.; Yavirach, A.; Rungsiyakull, C.; Rungsiyakull, P. The Integration of Gold Nanoparticles into Dental Biomaterials as a Novel Approach for Clinical Advancement: A Narrative Review. *J. Funct. Biomater.* **2024**, *15*, 291. [CrossRef] [PubMed]

Journal of
*Functional*
*Biomaterials*

MDPI

*Article*

# Reconstructive Approach in Residual Periodontal Pockets with Biofunctionalized Heterografts—A Retrospective Comparison of 12-Month Data from Three Centers

**Anton Friedmann [1,*], Pheline Liedloff [1], Meizi Eliezer [2], Arthur Brincat [3,4], Thomas Ostermann [5] and Daniel Diehl [1,6]**

1 Department of Periodontology, Faculty of Health, Witten/Herdecke University, Alfred-Herrhausen-Str. 50, 58455 Witten, Germany; pheline.liedloff@uni-wh.de (P.L.); daniel.diehl@uni-wh.de (D.D.)
2 Independent Researcher, Tel Aviv 3640577, Israel; meizi.eliezer@gmail.com
3 Independent Researcher, 83000 Toulon, France; arthurbrincat@gmail.com
4 Department of Periodontology, Service of Odontology, AP-HM, UFR of Odontology, Aix-Marseille University, 13005 Marseille, France
5 Department of Psychology, Faculty of Health, Witten/Herdecke University, 58455 Witten, Germany; thomas.ostermann@uni-wh.de
6 Institute of Pharmacology and Toxicology, Faculty of Health, Witten/Herdecke University, 58453 Witten, Germany
* Correspondence: anton.friedmann@uni-wh.de; Tel.: +49-2302926655

**Abstract:** The regenerative capacity of well-preserved blood clots may be enhanced by biologics like enamel matrix derivative (EMD). This retrospective analysis compares outcomes reported by three centers using different heterografts. Center 1 (C1) treated intrabony defects combining cross-linked high-molecular-weight hyaluronic acid (xHyA) with a xenograft; center 2 (C2) used EMD with an allograft combination to graft a residual pocket. Center 3 (C3) combined xHyA with the placement of a resorbable polymer membrane for defect cover. Clinical parameters, BoP reduction, and radiographically observed defect fill at 12-month examination are reported. The 12-month evaluation yielded significant improvements in PPD and CAL at each center ($p < 0.001$, respectively). Analyses of Covariance revealed significant improvements in all parameters, and a significantly greater CAL gain was revealed for C2 vs. C1 ($p = 0.006$). Radiographic defect fill presented significantly higher scores for C2 and C3 vs. C1 ($p = 0.003$ and $= 0.014$; C2 vs. C3 $p = 1.00$). Gingival recession increased in C1 and C3 ($p = 1.00$), while C2 reported no GR after 12 months (C2:C1 $p = 0.002$; C2:C3 $p = 0.005$). BoP tendency and pocket closure rate shared similar rates. Within the limitations of the study, a data comparison indicated that xHyA showed a similar capacity to enhance the regenerative response, as known for EMD. Radiographic follow-up underlined xHyA's unique role in new attachment formation.

**Keywords:** heterografts; EMD; xHyA; synthetic polymer barrier; bovine xenograft with hydroxyapatite; allograft

## 1. Introduction

Periodontitis is a chronic inflammatory disease caused by dysbiotic plaque biofilms that result in irreversible host-mediated damage to the tooth-supporting apparatus [1]. Depending on the rate of disease progression, periodontal bone loss may present itself as vertically configured, so-called intrabony, defects [2]. Regenerative surgery is the preferred method for addressing residual intrabony periodontal defects after non-surgical periodontal therapy, as recommended by the guidelines for treating periodontitis stages 1 to 3 [3]. In a systematic review and meta-analysis, regenerative strategies using enamel matrix derivative (EMD) or guided tissue regeneration (GTR) were found to be more effective than open-flap debridement (OFD) in terms of clinical attachment level (CAL) gain and reduction in probing depth (PD). These strategies showed a significant improvement in

CAL gain and PD reduction once implemented [4]. The overall superiority was expressed by a 1.27 mm greater CAL gain achieved with EMD and 1.43 mm achieved with GTR. Furthermore, this systematic review recommended the use of bone substitutes for intrabony defects with severely reduced bone walls to stabilize soft tissue and prevent collapse [2]. Moreover, space maintenance is crucial for both blood clot and tissue formation, according to established GTR principles [5,6]

A meta-analysis calculated the effect named "pocket closure" on behalf of 12 published randomized controlled trials (RCTs) addressing the efficacy of either GTR or EMD over OFD. The results revealed a 61.4% rate of closure looking at sites with $\leq$3 mm residual probing depth (PD), whereas a 92.1% closure rate was apparent once considering sites with a residual PD $\leq$ 4 mm after a 12-month post-op period [7].

In recent years, a new agent has proved sufficient in periodontal regeneration after its beneficial role in soft tissue healing had been demonstrated before. Studies conducted in vitro, pre-clinically, and as clinical case series documented the sufficient contribution of adjunctively applied hyaluronic acid to cell and tissue reactions. In particular, cross-linked high-molecular-weight hyaluronic acid (xHyA) showed a sufficient enhancement in soft tissue healing in donor sites for the retrieval of free gingival grafts from the palate [8]. The same formulation effectively supported soft tissue flap stabilization in recession coverage procedures carried out in an animal model as well as in patients [9,10]. The periodontal healing of surgically created intrabony defects was superior regarding a newly formed periodontal ligament and new cementum according to histomorphometric evaluation in a pre-clinical study by Shirakata et al. This group repeated the experiment by creating acute furcation grade 3 defects using the same dog model and confirmed the results from the previous study for the xHyA-treated defects [10,11]. A randomized clinical trial investigating three-wall intrabony defects further demonstrated the non-inferiority of xHyA-treated sites compared to EMD use for surgical regenerative treatment regarding CAL gain and PD reduction outcome after 24 months of follow-up [12]. Moreover, Bozic et al. achieved a >90% pocket closure rate by surgically applying xHyA with a porcine particulate xenograft after 12 months of healing [13].

Apart from clinical studies, the interaction between xHyA and fibroblasts derived from the periodontal ligament was elucidated by an in vitro experiment performed on dentin discs [14]. Another experimental study reported that the presence of xHyA on collagen substrates enhanced the gene transcription rate for bone-related proteins by osteoblast-like cells in vitro [15]. An in vitro study showing xHyA's impact on the transcription rate of the specific mRNAs encoding for cementoblast differentiation and on their enhanced proliferation was just released [16].

However, while the positive effects of the abovementioned biologics were evident, clinical studies comparing the effects contributed by EMD or xHyA to the regenerative surgical treatment of intrabony periodontal defects still need to be conducted. In this retrospective study, we investigated the outcomes of these two bioactive materials in combination with different adjunctive biomaterials for the surgical regenerative treatment of deep intrabony defects over 12 months.

## 2. Materials and Methods

The patients recruited were routinely treated periodontitis patients presenting with a diagnosis of stage 3 or 4 periodontitis regardless of their grading [17]. In all three centers, patients underwent a course of systematic subgingival instrumentation according to recommendations from the guidelines of the EFP concerning steps 1 and 2 before surgery scheduling [3]. The regenerative approach was favored once the re-evaluation values justified a step-3 surgical therapy of residual pockets, i.e., with a PPD exceeding 6 mm with or without BoP.

Principal investigators represented by A.B. for center 1 (C1), M.E. for center 2 (C2), and A.F. for center 3 (C3) were calibrated regarding the surgical technique applied, data evaluation, and inclusion criteria for approaching the residual pockets. All principal investigators were professionally trained periodontists with similar experience (three-year postgraduate degree, at least 10 years of practicing periodontics), and they performed all the surgeries. The concordant intention was to enhance tissue response to bone substitutes or membranes applied by combining them with bioactive formulations, either enamel matrix derivative (EMD) or high-molecular-weight crosslinked hyaluronic acid (xHyA), aiming at their biofunctionalization. The combination of grafting or membrane material with the bioactive molecules was addressed as a heterograft in each subgroup. Systemically healthy patients were included in this retrospective analysis by each center only. The modified papilla preservation incision design [18], full-flap elevation, and releasing incision for coronal flap advancement, as well as meticulous instrumentation of the root surface and thorough degranulation, were uniformly agreed for the surgical protocol. The protocol standardized neither the type of suture nor the suture technique. There were no restrictions regarding the route and type of instrumentation of the defects; i.e., ultrasonic or piezo devices were used as well as hand instruments. After completing thorough instrumentation, the defects received biomaterials considered supportive for regenerative healing. Each group was free to choose the biomaterial combination for regenerative surgery according to its own preference. The Ethics Committee of Witten/Herdecke University approved the retrospective analysis of the data set from the three centers (S-203/2021, amendment from 2023).

Each operator assessed clinical parameters (PPD, CAL, BoP, and GR for recession) by means of a manual periodontal probe on a regular basis during supportive periodontal therapy (SPT) visits. Clinically assessed values, as well as data regarding defect intrabony defect depth, defect angle, and defect wall number at baseline (prior to surgery) and after a period of 12 months, were reported. The radiographical findings were assessed on periapical 2D radiographs obtained digitally via the parallel technique using a sensor holder (Sidexis, Sirona, Bensheim, Germany) at baseline and 12 months post-op at each center. The calculation of tissue alterations revealed by comparison of both radiographs was reported by each center itself.

Center 1 used a combination of xHyA (HyaDent BG, Regedent AG, Zürich, Switzerland) and a collagen enhanced by hydroxyapatite particles (Collapat II, Symatese, Chaponost, France). Center 2 applied a combination of EMD (Emdogain, Straumann Group, Basel, Switzerland) and an allograft containing 50% cancellous and 50% cortical allograft bone (LifeNet Health, Virginia Beach, VA, USA). Both centers applied the materials to the intrabony pocket closing the site by coronally repositioning the soft tissue flap; neither center used a membrane. For the EMD application, the site was pre-conditioned by using 24% EDTA gel (Pref Gel, Straumann Group), taking care of the bloodless condition of the defect area prior to EMD gel application thereafter, according to recommendations from the manufacturer. The xHyA application also followed the manufacturer's recommendation; however, any pre-conditioning of the site was redundant and therefore omitted. The rehydration of either bone substitute was carried out on a tray before grafting the defect with a particulate heterograft. The rehydration afforded as much bioactive material as necessary to completely cover the total volume of the graft. The amount of EMD used per site amounted in total to one dose of 0.7 mL, while one ampule of xHyA contained 1.2 mL of the hyaluronic gel. An overview of the applied heterografts is shown in Table 1.

Center 3 used xHyA alone for filling the intrabony defect component, placing a polylactic poly-lactid polymer membrane (Guidor matrix barrier, Sunstar, Schönau, Germany) at the crest of the alveolar ridge before closing the site with a soft tissue flap in a similar way to both other centers. The membrane was rehydrated by xHyA similarly to the rehydration of the bone substitute in the other two centers.

The post-op regimen included pain medication, irrigation with CHX for a duration of 2 weeks, and local topical use of CHX gel for several weeks following suture removal. Each center was responsible for the choice of systemically administrated antibiotics for every case.

**Table 1.** Heterografts used in the study.

| | | | |
|---|---|---|---|
| C1 | xHyA | BDDE-crosslinked hyaluronic acid | HyaDent BG, Regedent, Zürich, Switzerland |
| | Collapat II | Bovine collagen + dispersed hydroxyapatite granules | Collapat II, Symatese, France |
| C2 | EMD | Enamel matrix derivative, Propylenglycolalginate (PGA), water | Emdogain, Straumann Group, USA |
| | OraGraft | Cortical/cancellous mineralized particulate 50/50 | LifeNet Health, USA |
| C3 | xHyA | BDDE-crosslinked hyaluronic acid | HyaDent BG, Regedent, Zürich, Switzerland |
| | Guidor matrix barrier | Polylactic polymer | Sunstar, Germany |

For metrical variables, e.g., PPD, CAL, and recession, descriptive statistics including mean, standard deviation, median, range, and percentages were applied to summarize the sample data. Differences between groups were calculated using Analysis of Covariance (ANCOVA) or a chi-square test in the case of nominal data with intraosseous depth, defect angle, and wall number as covariates. For pairwise group comparisons, the Bonferroni post hoc test was used. A two-tailed significance level of $\alpha = 5\%$ was applied for all analyses.

## 3. Results

All three centers recorded and reported uncomplicated healing. All patients were compliant with the SPT program and appeared at individual intervals for re-evaluation and cleaning visits. At one-year re-evaluation, all patients from three centers demonstrated significantly improved clinical parameters and positive alterations in crestal bone height when followed up radiographically.

C1 enrolled 18 patients with 19 treated defects, C2 accounted for 21 patients with an equal number of treated teeth, and C3 enrolled 15 patients with 15 teeth and sites to treat, respectively. The homogeneity in patient age and defect morphology included in the three centers was confirmed by non-significant differences in the defect angle, intrabony depth component, number of defect walls, initial probing depth (PD), and clinical attachment loss (CAL) loss. Furthermore, age, gender, and smoking habits were similarly distributed among the patients from each center (Table 2). The correlation between the outcome and the radiographic defect diminution (RDD/Δdefect fill) outcome was statistically significant only for the baseline value and the intraosseous defect component ($p < 0.001$); the initial number of bony walls ($p = 0.174$) and the defect angle ($p = 0.843$) were non-significant. Moreover, all groups exhibited a similar distribution of morphologic defect characteristics (Table 2).

While the PD reduction was similarly effective in all three centers (Tables 2 and 3), the inner group comparison revealed statistically significant differences in attachment-level gain reported by C1, C2, and C3 ($p < 0.001$, respectively) (Table 2; Figure 1). Figures 2–4 depict and illustrate one representative case per center including clinical images and periapical X-ray at baseline and 12 months post-op.

The ΔCAL comparison between centers favored center 2 vs. center 1 with a $p = 0.006$; the difference between C2 and C3 was statistically non-significant ($p = 0.718$). The radiographic bone fill was significantly greater in patients from centers 2 and 3 vs. center 1 ($p = 0.003$ and $= 0.014$, respectively) (Table 4). The difference in radiographically documented defect fill between C2 and C3 was statistically non-significant ($p = 1.0$). As corroborated by the 12-month results, both the significant clinical attachment gain and radiographic alveolar bone improvement remained constantly unaltered during the observation period (Table 2, Figures 1–3). The recession increased from baseline to the 12-month visit by 1.2–1.3 mm on average for center 1 and 3, while center 2 recorded a minimal recession increase of less than 0.5 mm.

**Table 2.** Patient demographics, pooled defect characteristics, and allocation per center (C1–C3).

| | C1 (n = 19) | C2 (n = 21) | C3 (n = 16) | Total (n = 56) | *p*-Value |
|---|---|---|---|---|---|
| **Age (years)** | | | | | |
| Mean ± SD | 58.5 ± 9.2 | 46.6 ± 9.3 | 53.1 ± 13.9 | 52.5 ± 11.7 | |
| Median | 57 | 46 | 54 | 55 | 0.004 |
| Min | 36 | 32 | 20 | 20 | |
| Max | 75 | 65 | 75 | 75 | |
| **Gender** | | | | | |
| Male | 10 (52.6%) | 4 (19.0%) | 6 (18.75%) | 17 (30.4%) | 0.085 |
| Female | 9 (47.4%) | 17 (81.0%) | 10 (62.5%) | 36 (64.3%) | |
| **Smoker** | | | | | |
| Yes | 3 (15.8%) | 4 (19.0%) | 0 (0.0%) | 7 (12.5%) | 0.192 |
| No | 16 (84.2%) | 17 (81.0%) | 16 (100.0%) | 49 (87.5%) | |
| **Localization** | | | | | |
| Mandible | 13 (68.4%) | 11 (52.4%) | 11 (68.8%) | 35 (62.5%) | 0.480 |
| Maxilla | 6 (9.47%) | 10 (9.02%) | 5 (9.22%) | 21 (37.5%) | |
| **Walls (n=)** | | | | | |
| 1 | 4 (21.1%) | 1 (4.8%) | 6 (37.5%) | 11 (19.6%) | |
| 2 | 11 (57.8%) | 17 (81.0%) | 8 (50.0%) | 36 (64.3%) | 0.137 |
| 3 | 4 (21.1%) | 3 (14.3%) | 2 (12.5%) | 9 (16.1%) | |
| **Intraosseous depth (mm)** | | | | | |
| Mean ± SD | 5.96 ± 1.68 | 5.70 ± 3.50 | 7.23 ± 2.44 | 6.22 ± 2.73 | |
| Median | 5.9 | 5.1 | 7.8 | 6.0 | 0.210 |
| Min | 3.3 | 1.9 | 2.4 | 1.9 | |
| Max | 10.6 | 15.3 | 11.0 | 15.3 | |
| **Defect angle (°)** | | | | | |
| Mean ± SD | 28.54 ± 9.80 | 31.32 ± 9.66 | 32.88 ± 14.22 | 30.82 ± 11.12 | |
| Median | 29.6 | 32.4 | 31.35 | 31.7 | 0.508 |
| Min | 14.0 | 14.6 | 16.5 | 14.0 | |
| Max | 50.5 | 44.2 | 64.5 | 64.5 | |
| **Defect width (mm)** | | | | | |
| Mean ± SD | 2.81 ± 099 | 2.55 ± 0.86 | N/A | 2.62 ± 0.92 | |
| Median | 2.7 | 2.4 | | 2.55 | 0.375 |
| Min | 1.5 | 1.1 | | 1.1 | |
| Max | 4.8 | 4.5 | | 4.8 | |
| **Antibiotics** | | | | | |
| Duration (n) | 7 days (19) | -- | 10 days (16) | | |
| Type (mg) | Amoxicillin (2000) | -- | Doxycycline (200) | | |
| **Analgesics** | | | | | |
| Duration (n) | If required | If required | If required | | |
| Type (mg) | Prednisone + Paracetamol (80 + 1000) | Ibuprofen (400) | Ibuprofen (600) | | |

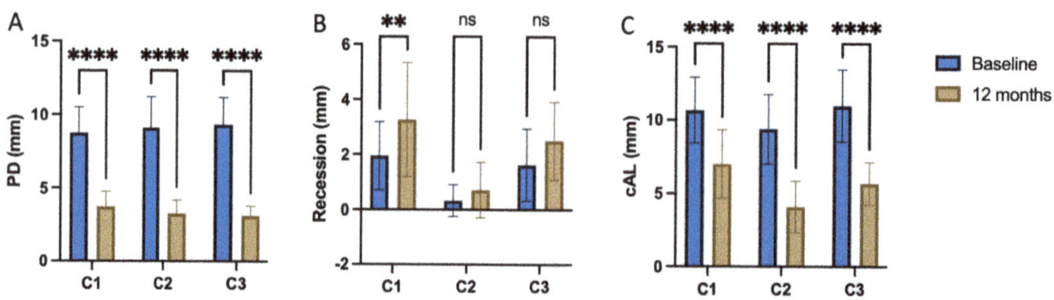


**Figure 1.** Intragroup comparisons between preoperative and post-operative clinical parameters (**A**) probing depth, (**B**) Recession and (**C**) clinical attachment loss. ** $p < 0.01$; **** $p < 0.0001$; ns—non-significant.

**Table 3.** Center-allocated change between baseline and 12-month exam for clinical parameters.

| | C1 | | C2 | | C3 | |
|---|---|---|---|---|---|---|
| | **Baseline** | **12-mo** | **Baseline** | **12-mo** | **Baseline** | **12-mo** |
| **PPD (mm)** | | | | | | |
| Mean ± SD | 8.74 ± 1.82 | 3.74 ± 1.05 | 9.29 ± 2.13 | 3.38 ± 0.92 | 9.50 ± 1.86 | 3.19 ± 0.66 |
| Median | 8.00 | 4.00 | 9.00 | 3.00 | 9.50 | 3.00 |
| Minimum | 7 | 2 | 7 | 2 | 6 | 2 |
| Maximum | 13 | 6 | 12 | 6 | 12 | 4 |
| *p*-value | <0.001 | | <0.001 | | <0.001 | |
| **CAL (mm)** | | | | | | |
| Mean ± SD | 10.68 ± 2.31 | 7.00 ± 2.29 | 9.62 ± 2.38 | 4.10 ± 1.76 | 11.25 ± 2.46 | 5.69 ± 1.45 |
| Median | 10.00 | 7.00 | 9.00 | 4.00 | 11.00 | 6.00 |
| Minimum | 7 | 4 | 7 | 2 | 7 | 3 |
| Maximum | 16 | 15 | 16 | 9 | 15 | 8 |
| *p*-value | <0.001 | | <0.001 | | <0.001 | |
| **REC (mm)** | | | | | | |
| Mean ± SD | 1.95 ± 1.27 | 3.26 ± 2.13 | 0.33 ± 0.58 | 0.71 ± 1.10 | 1.62 ± 1.31 | 2.50 ± 1.41 |
| Median | 2.00 | 3.00 | 0.00 | 0.00 | 1.00 | 3.00 |
| Minimum | 0 | 0 | 0 | 0 | 0 | 0 |
| Maximum | 5 | 10 | 2 | 3 | 4 | 5 |
| *p*-value | 0.003 | | 0.008 | | 0.029 | |

a b c d e f

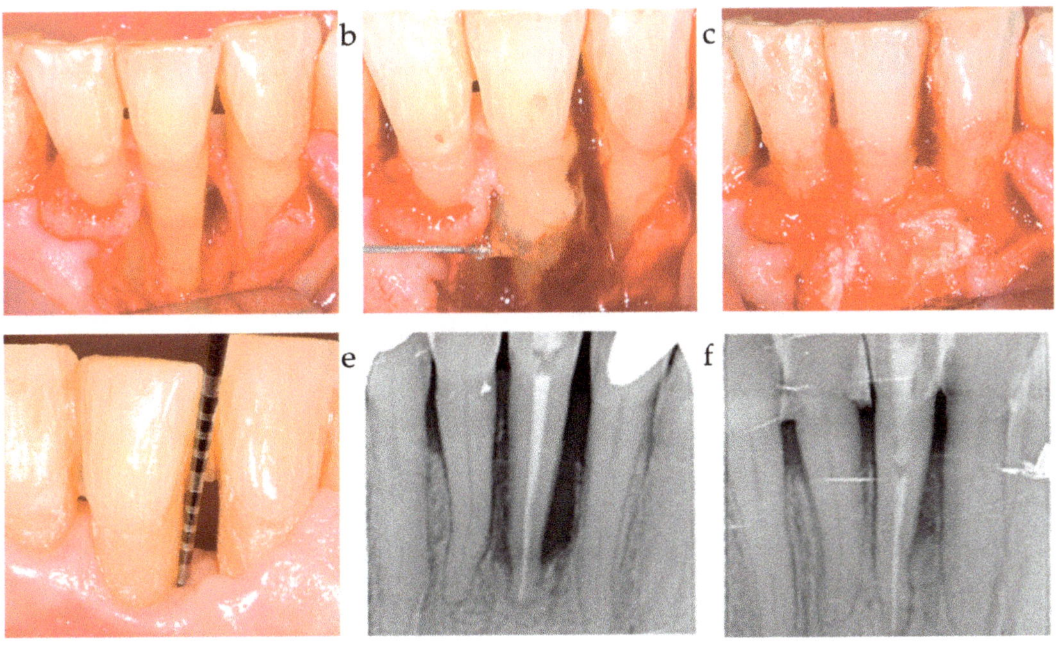

**Figure 2.** Center 1 showcase: (**a**,**b**) radiographically observed change in bone level around tooth 31 before surgery and 12 months post-op; (**c**,**d**) defect extension and defect grafting at surgery; (**e**,**f**) the result of grafting.

The rate for pocket closure was estimated at >90% in all treated sites regardless of the type of biomaterial. In detail, looking at the residual PD ≤ 4 mm without BoP, C1 showed 89.5%, C2 showed 95.3%, and C3 showed 93.4% pocket closure rates at the level of an residual probing depth of <4 mm without bleeding on probing. BoP appeared sufficiently reduced in all treated defects at an overall rate of 93%, without a great difference between three centers.

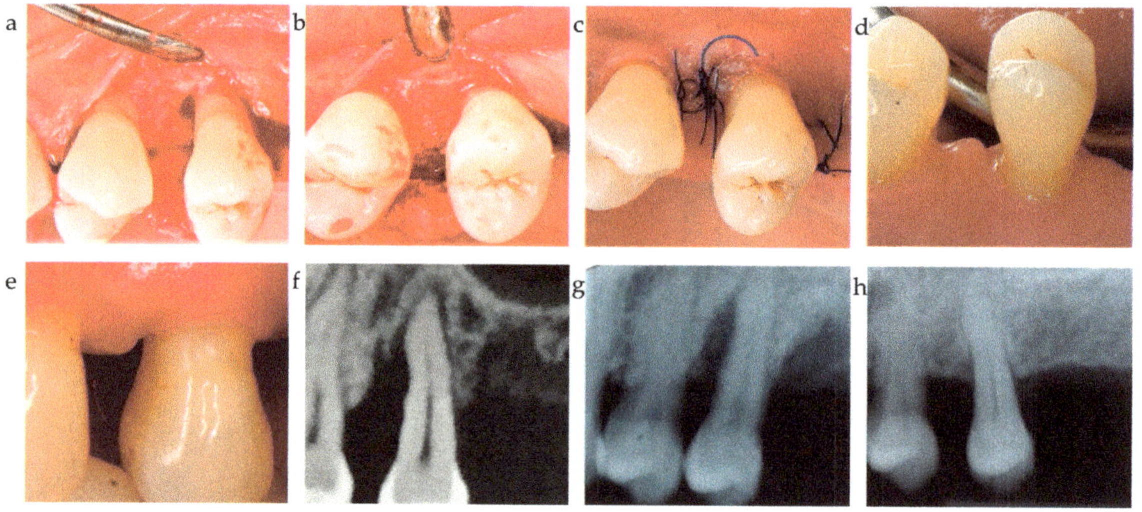

**Figure 3.** Center 2 showcase: (**a–c**) defect extension, defect grafting at surgery, and suture; (**d**,**e**) clinical outcome at 12-month exam. (**f–h**) radiographically observed change in bone level around tooth 25 before surgery and 12 months post-op.

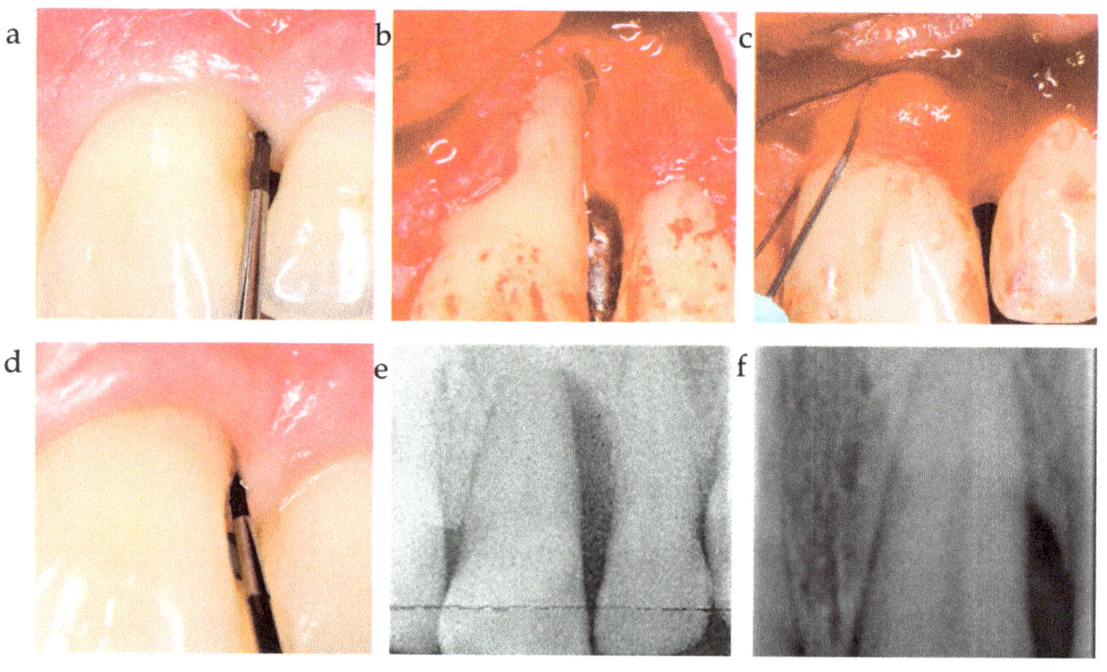

**Figure 4.** Center 3 showcase: (**a**,**b**) PPD and defect extension; (**c**,**d**) defect grafting at surgery as result of barrier placement and xHya application, and clinical outcome at 12-month exam. (**e**,**f**) radiographically observed change in bone level around tooth 21 before surgery and 12 months post-op.

**Table 4.** Comparison of Δ-values between centers and covariate analyses.

| | C1 (n = 19) | C2 (n = 21) | C3 (n = 16) | *p*-Value | | | | Significant Covariates |
|---|---|---|---|---|---|---|---|---|
| | | | | Overall | C1 vs. C2 | C2 vs. C3 | C1 vs. C3 | |
| **ΔPPD** | 4.95 ± 1.71 | 5.81 ± 1.78 | 6.25 ± 1.88 | 0.192 | 0.287 | 1.00 | 0.476 | Intraosseous depth ($p < 0.001$) |
| **ΔCAL** | 3.68 ± 1.67 | 5.86 ± 2.37 | 5.53 ± 1.92 | 0.007 | 0.006 | 0.718 | 0.158 | Intraosseous depth ($p < 0.001$) |
| **REC** | 1.32 ± 1.67 | 0.04± 0.01 | 1.33 ± 1.11 | 0.015 | 0.031 | 0.038 | 1.00 | - |
| **Δdefect fill** | 3.33 ± 1.76 | 4.95 ± 2.43 | 5.97 ± 2.49 | 0.002 | 0.003 | 1.00 | 0.014 | Intraosseous depth ($p < 0.001$) |

Among the covariates, the intraosseous defect depth tested significant for the outcome in ΔPPD, ΔCAL and Δdefect fill ($p < 0.001$). Other covariates, such as defect angle, number of defect walls, localization, or defect width (if reported), were not significantly associated with the clinical outcome (Table 4).

## 4. Discussion

This retrospective data analysis displayed a significant improvement for the patients treated by any of the three centers. Each center achieved a significant reduction in PPD at a clinically relevant level, with pocket closure rates ranging from 90 to 94%. This change was accompanied by significant attachment-level gains in each group. The results are in line with a recent systematic review published by Nibali et al., who reported an adjunctive benefit of regenerative procedures compared with open-flap debridement alone [4]. The authors also emphasized that the addition of deproteinized bovine bone mineral may further improve the outcome of GTR. However, in this retrospective analysis, the clinical attachment gain of C3 was equal to or even superior to the groups that used particulate bone substitutes. This may be rooted in the application of hyaluronic acid, which provides extended blood clot stability and a significant increase in osteogenesis [19,20].

To date, various in vitro studies have validated the beneficial mode of action provided by hyaluronic acid in periodontal tissue regeneration. Via binding to its canonical receptors CD44 and RHAMM, it has been shown to increase proliferation, migration, and cell metabolism in periodontal ligament fibroblasts (PDLs), mesenchymal stromal cells (MSCs), and cementoblasts. In osteoblasts and cementoblasts, xHyA may also stimulate the expression of bone-specific genes, while it shifts macrophage polarization towards an anti-inflammatory M2 phenotype (Figure 5) [14,16,21–24]. In a recent histomorphometric study in dogs, the combination of a resorbable matrix and xHyA was also shown to be superior, underlining the regenerative capacity of this biomaterial [25]. Moreover, other clinical studies reported that the adjunctive application of hyaluronic acid may improve CAL gain by a significant margin [26].

The regenerative potential of EMD in the formation of new attachment has been confirmed by a plethora of RCTs and human histological as well as in vitro studies (Figure 5) [27–33]. Quite expectedly, the results of C2 are, therefore, in the range of the latest clinical trials investigating EMD heterografts [34,35]. However, a comparison of our study results with the literature should be made with reasonable caution since the presented study was neither randomized nor controlled.

Nevertheless, the effectiveness of the chosen material composition proved to result in statistically significant differences from the baseline outcomes each center reported. Looking at the baseline number of bony walls, which were almost alike in all three centers ($p = 0.137$), the results from center 3 reported for the first time a significant CAL and bone gain accompanying xHyA use without a bone substitute in defects presenting a diminished number of bone walls (i.e., 1.5 on average). Stabilizing the defect via a polymer-derived membrane, the pocket closure effect at the level of a residual 3 mm probing depth was constantly observed after 12 months (93%).

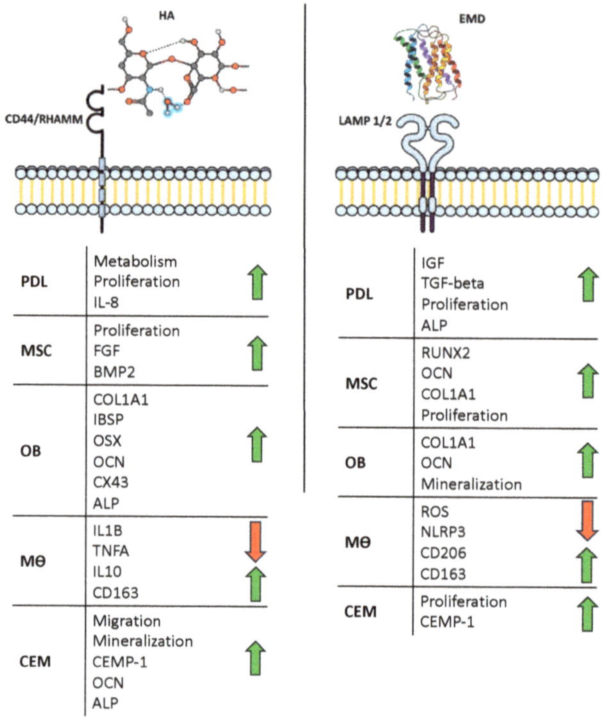


**Figure 5.** Schematic illustration of the mechanism of action reported for xHya (left) and EMD (right). PDL = periodontal ligament fibroblast; MSC = mesenchymal stromal cell; OB = osteoblast; MO = macrophage; CEM = cementoblast. Red arrows indicate downregulation, green arrows indicate upregulation.

The EFP guidelines recommend that in the case of a minimized number of bony walls, combined use with a particulate bone graft prevents the risk of tissue collapse into the defect. Thus, centers 1 and 2 combined the use of biologics (xHyA, EMD) with either a xenograft (C1) or an allograft (C2) according to the recommendations for treating defects characterized by a diminished number of walls [36,37]. Since C3 yielded significantly higher CAL gain values, the inconsistency with C1 may have rather been rooted in the choice of the substitute material. The bone substitute was a fleece with 98.9% porosity that has been proposed for tissue engineering [38]. The xenograft was recommended as a hemostatic device for enclosed defects [39]. Used in this series as a graft in an open periodontal pocket environment, even in combination with xHyA, the material may have undergone more rapid degradation compared with the allograft used by C2. The high proportion of collagen in this biomaterial may have been responsible for a rapid resorption accompanied by a partial collapse of the flap into the intraosseous defect. This healing pattern was then associated with an increasing post-operative recession and reduced the potential for the regeneration of the intraosseous component. Previous studies have pointed out that the type of collagen may be crucial for supporting the bone regeneration process and fast collagen degradation may be associated with limited outcomes [40].

Center 3, on the other hand, did not use any particulate material to avoid an artificial radio-opacity in the defect area and applied a polymer barrier membrane instead. This polymer membrane was shown to significantly improve clinical attachment levels in a variety of clinical studies [41]. In a 6-year observational study, Stavropoulos and Karring showed an attachment-level gain with a PPD improvement of $3.8 \pm 1.1$ mm and a mean CAL gain of $3.8 \pm 1.4$ mm observed after 1 year. Within this frame of reference, it appears reasonable to suggest that the inconsistency between C1 and C3 was indeed rooted in the choice of substitute material and the fact that xHya was demonstrated to be an efficient addition to the GTR technique applied by center 3 [42].

From this standpoint, it is rational to suggest that more research should focus on the substitute material within a functionalized heterograft, as EMD and xHya already exhibit sufficient evidence. Unfortunately, the membrane material used by C3 in combination with xHyA has been withdrawn from the market by the manufacturer due to recent legacy MDR regulations for European countries. Thus, making a warranted reproduction in a prospective study design impossible for the moment. Nevertheless, further studies comparing heterografts and guided tissue regeneration with the same biological agent would be meaningful.

All patients treated in this retrospective series were systemically healthy to exclude potential confounders. Nevertheless, the centers were not calibrated in terms of post-operative pain medication or antibiotics. As indicated by Table 2, different analgesic medications were applied throughout the healing period if required. Non-steroidal anti-inflammatory drugs (NSAIDs) have been proposed as an adjunct to periodontal treatment repeatedly, for obvious reasons [43]. Being cyclooxygenase inhibitors, they have been shown to reduce periodontal inflammation substantially [44,45]. However, adjunctive effects of reasonable dosing schemes on periodontal therapy were not confirmed by clinical studies [46]. Bearing in mind that the actual usage of NSAIDs was even less prominent in this study, it appears quite unlikely that the analgesic medication may pose a source of bias in this analysis.

Regarding antibiotics, a recent systematic review and meta-analysis evaluated the possible adjunctive effects of antimicrobial drugs on periodontal regenerative surgery. From a total of 105 randomized clinical trials, the authors were unable to detect any additional benefits to the treatments in terms of CAL gain, indicating that the varying antibiotics prescribed in this study had no significant impact on the clinical or radiographic outcomes [47].

Taken together, all heterografts yielded significant improvements in a range that is expected from the clinical literature. However, owing to the retrospective nature, this study has some limitations. Since no randomization was applied and no controls were included, intergroup comparisons should be regarded with caution. However, one must bear in mind that extensive trials comparing both EMD and xHya in a randomized setting are not available. Moreover, the clinical literature provides sufficient evidence for the regenerative efficacy of both biomaterials, rendering resource-intensive clinical studies for the purpose of identifying a superior biologic at least questionable. Also, measurable differences would possibly just be found regarding the handling of the materials in a heterograft and not the clinical outcome, since EMD usually requires defect surfaces free of blood. This, however, remains a matter of speculation.

In all three centers, as well as in the relevant literature, the CAL gain and defect fill were subject to quite high standard deviations. While this is a natural occurrence in clinical studies that can be dealt with by an adequately powered study design, it remains intriguing to understand whether patients exhibited interindividual differences in susceptibility for a specific biomaterial. For instance, the HA-binding receptor CD44 naturally expresses differing transcript variants, one of which lacks HA-binding ability [27,48]. An overexpression of this specific variant in the periodontal tissues may thus lead to a diminished efficacy of xHya-functionalized grafts. Identifying individual markers for biomaterial susceptibility may provide clinicians with an evidence-based toolkit for targeted regenerative therapies in the future.

## 5. Conclusions

In summary, this retrospective analysis of biofunctionalized heterografts across three centers demonstrated significant improvements in pocket-probing depth reduction and attachment-level gains. Center 3 showcased noteworthy clinical attachment-level improvements without using particulate bone substitutes, emphasizing the potential benefits of hyaluronic acid application in GTR. While the study's limitations are acknowledged, the findings highlight the effectiveness of biologics like EMD and hyaluronic acid in periodontal regenerative surgery, prompting the need for future randomized controlled trials to optimize treatment combinations for enhanced patient outcomes.

**Author Contributions:** Conceptualization, A.F. and D.D.; methodology, A.B., M.E. and A.F.; software, T.O.; validation, P.L.; formal analysis, P.L.; data curation, T.O.; writing—original draft preparation, A.F. and D.D.; writing—review and editing, A.B., M.E., P.L. and T.O. All authors have read and agreed to the published version of the manuscript.

**Funding:** This research received no external funding.

**Institutional Review Board Statement:** The study was conducted in accordance with the Declaration of Helsinki, and approved by the Institutional Ethic Committee at Witten/Herdecke University, (protocol code S-203/2021 from 21 September 2021, amendment from 21 March 2023).

**Informed Consent Statement:** Informed consent was obtained from all subjects involved in the study.

**Data Availability Statement:** The data that support the findings of this study are available upon reasonable request from the corresponding author, A.F. The data are not publicly available due to ethical concerns regarding patient information.

**Conflicts of Interest:** The authors declare no conflicts of interest.

## References

1. Hajishengallis, G. Immunomicrobial pathogenesis of periodontitis: Keystones, pathobionts, and host response. *Trends Immunol.* **2014**, *35*, 3–11. [CrossRef] [PubMed]
2. Lang, N.P. Focus on intrabony defects–conservative therapy. *Periodontology* **2000**, *22*, 51–58. [CrossRef] [PubMed]
3. Sanz, M.; Herrera, D.; Kebschull, M.; Chapple, I.; Jepsen, S.; Berglundh, T.; Sculean, A.; Tonetti, M.S.; Participants, E.W.; Consultants, M.; et al. Treatment of stage I–III periodontitis—The EFP S3 level clinical practice guideline. *J. Clin. Periodontol.* **2020**, *47*, 4–60. [CrossRef] [PubMed]
4. Nibali, L.; Koidou, V.P.; Nieri, M.; Barbato, L.; Pagliaro, U.; Cairo, F. Regenerative surgery versus access flap for the treatment of intra-bony periodontal defects: A systematic review and meta-analysis. *J. Clin. Periodontol.* **2020**, *47*, 320–351. [CrossRef] [PubMed]
5. Wikesjö, U.M.; Qahash, M.; Thomson, R.C.; Cook, A.D.; Rohrer, M.D.; Wozney, J.M.; Hardwick, W.R. Space-providing expanded polytetrafluoroethylene devices define alveolar augmentation at dental implants induced by recombinant human bone morphogenetic protein 2 in an absorbable collagen sponge carrier. *Clin. Implant Dent. Relat. Res.* **2003**, *5*, 112–123. [CrossRef] [PubMed]
6. Wikesjö, U.M.; Lim, W.H.; Razi, S.S.; Sigurdsson, T.J.; Lee, M.B.; Tatakis, D.N.; Hardwick, W.R. Periodontal repair in dogs: A bioabsorbable calcium carbonate coral implant enhances space provision for alveolar bone regeneration in conjunction with guided tissue regeneration. *J. Periodontol.* **2003**, *74*, 957–964. [CrossRef]
7. Aimetti, M.; Fratini, A.; Manavella, V.; Giraudi, M.; Citterio, F.; Ferrarotti, F.; Mariani, G.M.; Cairo, F.; Baima, G.; Romano, F. Pocket resolution in regenerative treatment of intrabony defects with papilla preservation techniques: A systematic review and meta-analysis of randomized clinical trials. *J. Clin. Periodontol.* **2021**, *48*, 843–858. [CrossRef]
8. Yıldırım, S.; Özener, H.Ö.; Doğan, B.; Kuru, B. Effect of topically applied hyaluronic acid on pain and palatal epithelial wound healing: An examiner-masked, randomized, controlled clinical trial. *J. Periodontol.* **2018**, *89*, 36–45. [CrossRef]
9. Pilloni, A.; Schmidlin, P.R.; Sahrmann, P.; Sculean, A.; Rojas, M.A. Correction to: Effectiveness of adjunctive hyaluronic acid application in coronally advanced flap in Miller class I single gingival recession sites: A randomized controlled clinical trial. *Clin. Oral Investig.* **2018**, *22*, 2961–2962. [CrossRef]
10. Shirakata, Y.; Nakamura, T.; Kawakami, Y.; Imafuji, T.; Shinohara, Y.; Noguchi, K.; Sculean, A. Healing of buccal gingival recessions following treatment with coronally advanced flap alone or combined with a cross-linked hyaluronic acid gel. An experimental study in dogs. *J. Clin. Periodontol.* **2021**, *48*, 570–580. [CrossRef]
11. Shirakata, Y.; Imafuji, T.; Nakamura, T.; Shinohara, Y.; Iwata, M.; Setoguchi, F.; Noguchi, K.; Sculean, A.; Dent, M. Cross-linked hyaluronic acid-gel with or without a collagen matrix in the treatment of class III furcation defects: A histologic and histomorphometric study in dogs. *J. Clin. Periodontol.* **2022**, *49*, 1079–1089. [CrossRef]
12. Pilloni, A.; Zeza, B.; Kuis, D.; Vrazic, D.; Domic, T.; Olszewska-Czyz, I.; Popova, C.; Kotsilkov, K.; Firkova, E.; Dermendzieva, Y. Treatment of Residual Periodontal Pockets Using a Hyaluronic Acid-Based Gel: A 12 Month Multicenter Randomized Triple-Blinded Clinical Trial. *Antibiotics* **2021**, *10*, 924. [CrossRef] [PubMed]
13. Božić, D.; Ćatović, I.; Badovinac, A.; Musić, L.; Par, M.; Sculean, A. Treatment of intrabony defects with a combination of hyaluronic acid and deproteinized porcine bone mineral. *Materials* **2021**, *14*, 6795. [CrossRef] [PubMed]

14. Fujioka-Kobayashi, M.; Müller, H.-D.; Mueller, A.; Lussi, A.; Sculean, A.; Schmidlin, P.R.; Miron, R.J. In vitro effects of hyaluronic acid on human periodontal ligament cells. *BMC Oral Health* **2017**, *17*, 44. [CrossRef]
15. Nobis, B.; Ostermann, T.; Weiler, J.; Dittmar, T.; Friedmann, A. Impact of cross-linked hyaluronic acid on osteogenic differentiation of SAOS-2 cells in an air-lift model. *Materials* **2022**, *15*, 6528. [CrossRef]
16. Hakki, S.S.; Bozkurt, S.B.; Sculean, A.; Božić, D. Hyaluronic acid enhances cell migration, viability, and mineralized tissue-specific genes in cementoblasts. *J. Periodontal Res.* **2023**. [CrossRef]
17. Caton, J.G.; Armitage, G.; Berglundh, T.; Chapple, I.L.C.; Jepsen, S.; Kornman, K.S.; Mealey, B.L.; Papapanou, P.N.; Sanz, M.; Tonetti, M.S. A new classification scheme for periodontal and peri-implant diseases and conditions–Introduction and key changes from the 1999 classification. *J. Clin. Periodontol* **2018**. [CrossRef]
18. Cortellini, P.; Tonetti, M.S. Clinical concepts for regenerative therapy in intrabony defects. *Periodontology* **2015**, *68*, 282–307. [CrossRef]
19. Pilloni, A.; Bernard, G. The effect of hyaluronan on mouse intramembranous osteogenesis in vitro. *Cell Tissue Res.* **1998**, *294*, 323–333. [CrossRef]
20. Xing, F.; Zhou, C.; Hui, D.; Du, C.; Wu, L.; Wang, L.; Wang, W.; Pu, X.; Gu, L.; Liu, L.; et al. Hyaluronic acid as a bioactive component for bone tissue regeneration: Fabrication, modification, properties, and biological functions. *Nanotechnol. Rev.* **2020**, *9*, 1059–1079. [CrossRef]
21. Asparuhova, M.B.; Chappuis, V.; Stähli, A.; Buser, D.; Sculean, A. Role of hyaluronan in regulating self-renewal and osteogenic differentiation of mesenchymal stromal cells and pre-osteoblasts. *Clin. Oral Investig.* **2020**, *24*, 3923–3937. [CrossRef] [PubMed]
22. Frasheri, I.; Tsakiridou, N.D.; Hickel, R.; Folwaczny, M. The molecular weight of hyaluronic acid influences metabolic activity and osteogenic differentiation of periodontal ligament cells. *Clin. Oral Investig.* **2023**, *27*, 5905–5911. [CrossRef] [PubMed]
23. Kim, H.; Cha, J.; Jang, M.; Kim, P. Hyaluronic acid-based extracellular matrix triggers spontaneous M2-like polarity of monocyte/macrophage. *Biomater. Sci.* **2019**, *7*, 2264–2271. [CrossRef] [PubMed]
24. Mathews, S.; Mathew, S.A.; Gupta, P.K.; Bhonde, R.; Totey, S. Glycosaminoglycans enhance osteoblast differentiation of bone marrow derived human mesenchymal stem cells. *J. Tissue Eng. Regen. Med.* **2014**, *8*, 143–152. [CrossRef] [PubMed]
25. Shirakata, Y.; Imafuji, T.; Nakamura, T.; Kawakami, Y.; Shinohara, Y.; Noguchi, K.; Pilloni, A.; Sculean, A. Periodontal wound healing/regeneration of two-wall intrabony defects following reconstructive surgery with cross-linked hyaluronic acid-gel with or without a collagen matrix: A preclinical study in dogs. *Quintessence Int.* **2021**, *52*, 308–316. [PubMed]
26. Briguglio, F.; Briguglio, E.; Briguglio, R.; Cafiero, C.; Isola, G. Treatment of infrabony periodontal defects using a resorbable biopolymer of hyaluronic acid: A randomized clinical trial. *Quintessence Int.* **2013**, *44*, 231. [PubMed]
27. Amin, H.D.; Olsen, I.; Knowles, J.; Dard, M.; Donos, N. Interaction of enamel matrix proteins with human periodontal ligament cells. *Clin. Oral Investig.* **2016**, *20*, 339–347. [CrossRef] [PubMed]
28. Cheng, L.; Li, Y.; Xia, Q.; Meng, M.; Ye, Z.; Tang, Z.; Feng, H.; Chen, X.; Chen, H.; Zeng, X. Enamel matrix derivative (EMD) enhances the osteogenic differentiation of bone marrow mesenchymal stem cells (BMSCs). *Bioengineered* **2021**, *12*, 7033–7045. [CrossRef]
29. Sordi, M.B.; Cabral da Cruz, A.C.; Panahipour, L.; Gruber, R. Enamel matrix derivative decreases pyroptosis-related genes in macrophages. *Int. J. Mol. Sci.* **2022**, *23*, 5078. [CrossRef]
30. Stout, B.M.; Alent, B.J.; Pedalino, P.; Holbrook, R.; Gluhak-Heinrich, J.; Cui, Y.; Harris, M.A.; Gemperli, A.C.; Cochran, D.L.; Deas, D.E. Enamel matrix derivative: Protein components and osteoinductive properties. *J. Periodontol.* **2014**, *85*, e9–e17. [CrossRef]
31. Sanz, M.; Tonetti, M.S.; Zabalegui, I.; Sicilia, A.; Blanco, J.; Rebelo, H.; Rasperini, G.; Merli, M.; Cortellini, P.; Suvan, J.E. Treatment of intrabony defects with enamel matrix proteins or barrier membranes: Results from a multicenter practice-based clinical trial. *J. Periodontol.* **2004**, *75*, 726–733. [CrossRef]
32. Sculean, A.; Kiss, A.; Miliauskaite, A.; Schwarz, F.; Arweiler, N.B.; Hannig, M. Ten-year results following treatment of intra-bony defects with enamel matrix proteins and guided tissue regeneration. *J. Clin. Periodontol.* **2008**, *35*, 817–824. [CrossRef] [PubMed]
33. Sculean, A.; Windisch, P.; Szendröi-Kiss, D.; Horváth, A.; Rosta, P.; Becker, J.; Gera, I.; Schwarz, F. Clinical and histologic evaluation of an enamel matrix derivative combined with a biphasic calcium phosphate for the treatment of human intrabony periodontal defects. *J. Periodontol.* **2008**, *79*, 1991–1999. [CrossRef] [PubMed]
34. Meyle, J.; Hoffmann, T.; Topoll, H.; Heinz, B.; Al-Machot, E.; Jervøe-Storm, P.M.; Meiß, C.; Eickholz, P.; Jepsen, S. A multi-centre randomized controlled clinical trial on the treatment of intra-bony defects with enamel matrix derivatives/synthetic bone graft or enamel matrix derivatives alone: Results after 12 months. *J. Clin. Periodontol.* **2011**, *38*, 652–660. [CrossRef] [PubMed]
35. Ogihara, S.; Tarnow, D.P. Efficacy of enamel matrix derivative with freeze-dried bone allograft or demineralized freeze-dried bone allograft in intrabony defects: A randomized trial. *J. Periodontol.* **2014**, *85*, 1351–1360. [CrossRef]
36. Matarasso, M.; Iorio-Siciliano, V.; Blasi, A.; Ramaglia, L.; Salvi, G.E.; Sculean, A. Enamel matrix derivative and bone grafts for periodontal regeneration of intrabony defects. A systematic review and meta-analysis. *Clin. Oral Investig.* **2015**, *19*, 1581–1593. [CrossRef]
37. Sculean, A.; Nikolidakis, D.; Schwarz, F. Regeneration of periodontal tissues: Combinations of barrier membranes and grafting materials–biological foundation and preclinical evidence: A systematic review. *J. Clin. Periodontol.* **2008**, *35*, 106–116. [CrossRef]
38. Basha, R.Y.; TS, S.K.; Doble, M. Design of biocomposite materials for bone tissue regeneration. *Mater. Sci. Eng. C* **2015**, *57*, 452–463. [CrossRef]

39. Kurien, T.; Pearson, R.; Scammell, B. Bone graft substitutes currently available in orthopaedic practice: The evidence for their use. *Bone Jt. J.* **2013**, *95*, 583–597. [CrossRef]
40. Friedmann, A.; Fickl, S.; Fischer, K.R.; Dalloul, M.; Goetz, W.; Kauffmann, F. Horizontal augmentation of chronic mandibular defects by the Guided Bone Regeneration approach: A randomized study in dogs. *Materials* **2021**, *15*, 238. [CrossRef]
41. Falk, H.; Laurell, L.; Ravald, N.; Teiwik, A.; Persson, R. Guided tissue regeneration therapy of 203 consecutively treated intrabony defects using a bioabsorbable matrix barrier. Clinical and radiographic findings. *J. Periodontol.* **1997**, *68*, 571–581. [CrossRef]
42. Stavropoulos, A.; Karring, T. Long-term stability of periodontal conditions achieved following guided tissue regeneration with bioresorbable membranes: Case series results after 6–7 years. *J. Clin. Periodontol.* **2004**, *31*, 939–944. [CrossRef] [PubMed]
43. Ren, J.; Fok, M.R.; Zhang, Y.; Han, B.; Lin, Y. The role of non-steroidal anti-inflammatory drugs as adjuncts to periodontal treatment and in periodontal regeneration. *J. Transl. Med.* **2023**, *21*, 149. [CrossRef] [PubMed]
44. Weaks-Dybvig, M.; Sanavi, F.; Zander, H.; Rifkin, B.R. The effect of indomethacin on alveolar bone loss in experimental periodontitis. *J. Periodontal Res.* **1982**, *17*, 90–100. [CrossRef] [PubMed]
45. Williams, R.; Jeffcoat, M.; Kaplan, M.; Goldhaber, P.; Johnson, H.; Wechter, W. Flurbiprofen: A potent inhibitor of alveolar bone resorption in beagles. *Science* **1985**, *227*, 640–642. [CrossRef] [PubMed]
46. Kurtiş, B.; Tüter, G.; Serdar, M.; Pınar, S.; Demirel, İ.; Toyman, U. Gingival crevicular fluid prostaglandin E2 and thiobarbituric acid reactive substance levels in smokers and non-smokers with chronic periodontitis following phase I periodontal therapy and adjunctive use of flurbiprofen. *J. Periodontol.* **2007**, *78*, 104–111. [CrossRef] [PubMed]
47. Nibali, L.; Buti, J.; Barbato, L.; Cairo, F.; Graziani, F.; Jepsen, S. Adjunctive effect of systemic antibiotics in regenerative/reconstructive periodontal surgery—A systematic review with meta-analysis. *Antibiotics* **2021**, *11*, 8. [CrossRef]
48. Nagano, O.; Saya, H. Mechanism and biological significance of CD44 cleavage. *Cancer Sci.* **2004**, *95*, 930–935. [CrossRef]

Journal of
*Functional*
*Biomaterials*

MDPI

*Article*

# In Vitro Effects of *Weissella cibaria* CMU and CMS1 on Receptor Activator of NF-κB Ligand (RANKL)-Induced Osteoclast Differentiation

**Geun-Yeong Park, Jeong-Ae Park and Mi-Sun Kang ***

R&D Center, OraTicx, Inc., Seoul 04782, Republic of Korea; gypark@oraticx.com (G.-Y.P.); pja@oraticx.com (J.-A.P.)
* Correspondence: jieenkang@oraticx.com; Tel.: +82-2-2138-2572

**Abstract:** Excessive osteoclast activity can promote periodontitis-associated bone destruction. The inhibitory mechanisms of *Weissella cibaria* strains CMU and CMS1 against periodontitis have not yet been fully elucidated. In this study, we aimed to investigate whether heat-killed (HK) *W. cibaria* CMU and CMS1 or their respective cell-free supernatants (CFSs) inhibit osteoclast differentiation and bone resorption in response to receptor activator of nuclear factor kappa-B ligand (RANKL)-treated RAW 264.7 cells. TRAP (tartrate-resistant acid phosphatase) staining and bone resorption assays revealed that both HK bacteria and CFSs significantly suppressed the number of TRAP-positive cells, TRAP activity, and bone pit formation compared to the RANKL-treated control ($p < 0.05$). HK bacteria dose-dependently inhibited osteoclastogenesis while selectively regulating certain genes in CFSs ($p < 0.05$). We found that disrupting the direct interaction between HK bacteria and RAW 264.7 cells abolished the inhibitory effect of HK bacteria on the expression of osteoclastogenesis-associated proteins (c-Fos, nuclear factor of activated T cells c1 (NFATc1), and cathepsin K). These results suggest that dead bacteria suppress osteoclast differentiation more effectively than the metabolites and may serve as beneficial agents in preventing periodontitis by inhibiting osteoclast differentiation via direct interaction with cells.

**Keywords:** *Weissella cibaria*; RAW 264.7 cells; osteoclast; bone resorption; periodontitis

## 1. Introduction

Periodontal disease is a prevalent and widespread chronic inflammatory disease that affects approximately 20–50% of the global population and encompasses gingivitis and periodontitis [1,2]. Moreover, periodontitis is a major cause of tooth loss. Upon infection with periodontal bacteria, pro-inflammatory cytokines increase, and inflammation occurs around the gums. The inflammatory response activates osteoclasts and destroys the alveolar bone near the gums, thereby compromising the capacity of the bone to support teeth, leading to tooth loss [3].

Osteoclasts regulate bone metabolism together with osteoblasts, another type of bone cell [4–6]. Osteoclasts and osteoblasts participate in bone matrix formation, and when periodontitis occurs, osteoclasts can be activated to resorb and decompose the bone. Receptor activator of nuclear factor (NF)-κB ligand (RANKL) is a cell signaling protein found in osteoblasts [7]. RANKL binds to RANK present in osteoclast precursors, activating the mitogen-activated protein kinase (MAPK) and NF-κB signaling pathways and sequentially activating nuclear factor of activated T cells c1 (NFATc1), which is essential for osteoclast differentiation [8,9]. Furthermore, it is known to promote osteoclast differentiation and bone resorption by inducing the expression of several genes [9–13]. Therefore, inflammatory periodontal disease associated with osteoclasts can be alleviated by suppressing osteoclast differentiation [14].

Effective treatments for managing periodontal disease are limited, and researchers are continually developing treatments to prevent periodontitis alongside oral hygiene management. Recently, as the importance of disease prevention has been emphasized worldwide, probiotics have been proposed as an alternative for periodontal disease prevention [15–18]. Given the crucial role of the oral microbiota in regulating alveolar bone remodeling [15], oral probiotics have been explored as a potential therapeutic intervention to maintain alveolar bone homeostasis and prevent periodontitis. Representative studies have been conducted employing various probiotics, including *Bifidobacterium*, *Lactobacillus*, and *Weissella* [17–28].

*Weissella cibaria* strains CMU and CMS1 were isolated from the saliva of children with good oral health and Gram-positive lactic acid bacteria with a short rod-shaped morphology [19]. Both animal and clinical studies, including in vitro evaluations, indicate that these strains can inhibit the progression of periodontitis by regulating the inflammatory response of host cells [18–22]. Kim et al. [18] demonstrated that bone loss in periodontitis-induced mice was prevented by treatment with the oral probiotic *W. cibaria* CMU, highlighting its capacity to reduce inflammatory cytokines in gum tissue. In addition, it has been proven that *W. cibaria* CMU inhibits the formation of pro-inflammatory cytokines in human gingival fibroblasts caused by periodontal disease-causing bacteria such as *Fusobacterium nucleatum*, *Prevotella intermedia*, and *Porphyromonas gingivalis* [22]. However, the effects of *W. cibaria* CMU and CMS1 on RANKL-induced osteoclast differentiation remain unexplored.

The murine macrophage cell line RAW 264.7 stands out as the primary host cell line for studying osteoclast differentiation owing to its pronounced expression of RANK in response to RANKL [29,30]. Several studies have reported that various probiotic bacterial strains inhibit RANK-induced osteoclastogenesis in RAW 264.7 cells [23–25]. Therefore, in this study, we evaluated the effects of oral probiotics *W. cibaria* CMU and CMS1 on RANKL-stimulated osteoclast differentiation in RAW 264.7 cells in vitro by elucidating the molecular mechanisms underlying osteogenesis inhibition.

## 2. Materials and Methods

### *2.1. Cell Culture and Osteoclast Differentiation In Vitro*

The RAW 264.7 cell line (murine macrophage) was purchased from the Korean Cell Line Bank (KCLB, Seoul, Republic of Korea). Cells were grown in Dulbecco's Modified Eagle Medium (DMEM; Gibco, Thermo Fisher Scientific, Gaithersburg, MD, USA) supplemented with 10% heat-inactivated fetal bovine serum (FBS; Gibco) and 1% antibiotic–antimycotic solution (GenDepot, Katy, TX, USA) at 37 °C in a 5% $CO_2$ humidified atmosphere. Experiments were conducted on cell passages 2 to 10. The cells were subcultured and plated at 80% confluency. For osteoclast differentiation, alpha-minimal essential medium (α-MEM; Welgene Inc., Daegu, Republic of Korea) containing 10% FBS was treated with RANKL (100 ng/mL; Peprotech, Cranbury, NJ, USA). The medium was changed every alternate day throughout the culture period.

### *2.2. Preparation of Heat-Killed Bacteria and Cell-Free Supernatants*

*W. cibaria* CMU (oraCMU) and CMS1 (oraCMS1) were grown aerobically in DeMan, Rogosa, and Sharpe (MRS) broth (Difco, Detroit, MI, USA) at 37 °C for 16 h. To prepare the heat-killed (HK) bacteria, the cultures were centrifuged (5000× *g*, 10 min, 4 °C), the resulting pellets were washed twice with phosphate-buffered saline (PBS), resuspended in DMEM, and the concentration was adjusted to an optical density (OD) of 0.5 at 600 nm (approximately $5 \times 10^8$ CFU/mL). The bacteria were then exposed to heat (110 °C) for 10 min. Cell-free supernatants (CFSs) were prepared by centrifuging the culture for 24 h, followed by filtration (0.22 μm; JET BIOFIL, Guangzhou, China) to remove cells. The CFSs were lyophilized, resuspended in DMEM, and filtered again (0.45 μm; JET BIOFIL).

### *2.3. Cell Viability Assay*

A viability assay kit (Cellrix, MediFab, Seoul, Republic of Korea) was used to measure cell viability after treatment with test substances (HK-oraCMU and HK-oraCMS1 or CFS-oraCMU and CFS-oraCMS1). RAW 264.7 cells were seeded on 96-well plates

(1 × $10^4$ cells/well) and incubated at 37 °C for 16 h. The culture medium used to induce the differentiation of RAW 264.7 cells into osteoclasts were replaced with α-MEM supplemented with 10% FBS and incubated with RANKL at various concentrations of HK bacteria (multiplicity of infection (MOI) = 1, 10, 100, or 1000) or CFSs (0.25, 0.5, 1, or 2 mg/mL) for 2 d. Subsequently, the culture was removed and carefully replaced with fresh medium containing a water-soluble tetrazolium-8 (WST-8) salt solution and incubated for 4 h at 37 °C under 5% $CO_2$. Cell viability was measured at 450 nm using a microplate reader (VersaMax, Molecular Devices, San Jose, CA, USA) and expressed as a percentage relative to the untreated negative control.

### 2.4. Tartrate-Resistant Acid Phosphatase (TRAP) Staining and Activity

RAW 264.7 cells were seeded on a 96-well plate (1 × $10^4$ cells/well) for 16 h and incubated for an additional 5 days in an induction medium containing RANKL and various concentrations of the test substances. The induction medium was replaced every other day. TRAP staining and activity assays were performed according to the manufacturer's instructions (Cosmo Bio, Tokyo, Japan) to detect osteoclasts. Briefly, the culture medium was removed, and the cells were washed with PBS, and fixed with 10% formalin neutral buffer for 5 min at 25 °C. Each well was washed three times with distilled water (DW), and chromogenic substrate in tartrate-containing buffer was added to each well and incubated for 60 min at 37 °C. The wells were washed again three times with DW and dried at 25 °C for 2 h. Stained cells were observed under an inverted microscope (OLYMPUS CKX53, Olympus, Tokyo, Japan), and multinucleated cells containing three or more nuclei were identified as osteoclasts. To evaluate TRAP activity, 30 μL of the collected culture medium described above was transferred to a new well, mixed with 170 μL of chromogenic substrate in tartrate-containing buffer (Cosmo Bio), and incubated at 37 °C for 3 h. TRAP activity was assessed at 540 nm, and the results were expressed as a percentage of the control (RANKL-treated only).

### 2.5. Bone Resorption Assay

The effect of the test substances on osteoclast-mediated bone resorption was determined using dentin discs (Immunodiagnostic Systems, Boldon, UK). RAW 264.7 cells were seeded on a 96-well plate (5 × $10^3$ cells/well) for 16 h and incubated for an additional 5 days in the induction medium containing RANKL and various concentrations of the test substances. The induction medium was replaced every alternate day. The medium was removed, treated with 5% sodium hypochlorite for 5 min, and washed three times with DW. Toluidine blue solution (0.1%; Sigma-Aldrich, St. Louis, MO, USA) was added to each well for 3 min, washed again with DW three times, dried at 25 °C for 2 h, and observed under an inverted microscope at 400× magnification. Image J software Ver. 1.54 (National Institutes of Health, Washington, DC, USA) was used to measure the total pit area. The pit area was expressed as a value relative to that of the control (RANKL-treated only).

### 2.6. Reverse Transcription (RT)-Quantitative Polymerase Chain Reaction (qPCR)

RT-qPCR was performed to investigate the effect of the test substances on the expression of osteoclast differentiation-mediated genes. RAW 264.7 cells were seeded on a 6-well plate (1 × $10^5$ cells/well), and the following day, the cells were replaced with induction medium containing RANKL and various concentrations of the test substances and subsequently cultured for an additional 2 days. Total RNA was extracted, and RT-qPCR was performed on a Rotor Gene Q system (Qiagen, Hilden, Germany) using the PrimeScript RT kit (Takara Bio, Shiga, Japan) and the PowerUp SYBR Green PCR Master Mix (Applied Biosystems, Thermo Fisher Scientific) as previously described [22]. The primer sequences were as follows: mouse *cathepsin K* forward (F), 5′-GAAGAAGACTCACCAGAAGCAG-3′ and reverse (R), 5′-TCCAGG TTATGGGCAGAGATT-3′; mouse *c-Fos* F, 5′-CGGGTTTCAACGCCG ACTA-3′ and R, 5′-TGGCACTAGAGACGGACAGAT-3′; mouse *NFATc1* F, 5′-GGTGCTGTC TGG CCATAACT-3′ and R, 5′-GCGGAAAGGTGGTATCTCAA-3′; mouse *TRAP* F, 5′-

GACAAGAGGTTCCAGGAGACC-3′; and R, 5′-GGGCTGGGGAAGTTCCAG-3′; mouse *osteoclast associated Ig-like receptor* (*OSCAR*) F, 5′-CTGCTGGTAACGGATCAGCT CCCCAGA-3′ and R, 5′-CCAAGG AGCCAGAACCTTCGAAACT-3′; mouse *dendritic cell specific transmembrane protein* (*DC-STAMP*) F, 5′-CCAAGGAGTCGTCCATGATT-3′ and R, 5′-GGCTGCT TTGATCGTTTCTC-3′; mouse *glyceraldehyde 3-phosphate dehydrogenase* (*GAPDH*) F, 5′-AGGT CGGTGTGAACGGATTTG-3′ and R, 5′-TGTAGACCATGTAGTTGAGGTCA-3′. Relative mRNA expression values were obtained by the $2^{-\Delta\Delta CT}$ method, and relative gene expression was normalized to GAPDH expression.

### 2.7. Western Blotting Analyses

RAW 264.7 cells were seeded on a 6-well plate ($1 \times 10^5$ cells/well), and the following day, the cells were replaced with induction medium containing RANKL and various concentrations of the test substances and subsequently cultured for an additional 2 days. After cell seeding, a cell culture insert (pore size: 0.4 μm; SPL Life Sciences, Pocheon, Republic of Korea) was placed onto the relevant well, and HK *W. cibaria* was added onto the filter membrane. After incubation, cells were washed once with ice-cold Dulbecco's PBS (GenDepot), and proteins were extracted using an EzRIPA Lysis kit (ATTO, Tokyo, Japan), according to the manufacturer's protocol. Proteins (25 μg) were resolved by SDS–PAGE on 10% acrylamide gels and transferred onto polyvinylidene fluoride membranes (0.45 μm; Amersham, Cytiva, Marlborough, MA, USA). Membranes were blocked with 5% skim milk in Tris-buffered saline with 0.1% Tween 20 (GenDepot) for 1 h at 25 °C and incubated overnight at 4 °C with the following primary antibodies: c-Fos, NFATc1, cathepsin K, phospho-p44/42 MAPK (p-ERK), ERK, p-IκBα, IκBα, p-p38, p38, p-SAPK/JNK (p-JNK), JNK, and β-actin. Membranes were washed and incubated with horseradish peroxidase-conjugated secondary antibodies at 25 °C for 1 h. c-Fos and NFATc1 were purchased from Santa Cruz Biotechnology (Dallas, TX, USA) and BD Biosciences (San Jose, CA, USA), respectively. All other antibodies were purchased from Cell Signaling Technology (Danvers, MA, USA).

Each protein band was visualized using a West-Q Chemiluminescent Substrate Kit Plus (GenDepot) and imaged using a Chemiluminescence Imaging System (WSE-6200 LuminoGraph II, ATTO). Protein expression was quantified using CSAnalyzer4 version 2.4.5 (ATTO), and the relative expression was normalized to β-actin expression.

### 2.8. Statistical Analysis

The data were expressed as the mean ± SD of three independent experiments. SPSS Statistics version 21.0 for Windows (IBM, Armonk, NY, USA) was used for statistical analysis. To assess differences between group means, one-way analysis of variance (ANOVA) or Welch's ANOVA calculations with the Duncan multiple test or Games–Howell post-hoc test were performed. The statistical significance level of $p < 0.05$ was adopted.

## 3. Results

### 3.1. Cytotoxicity of W. cibaria CMU and CMS1 in RAW 264.7 Cells

During the differentiation process into RANKL-treated osteoclasts, high concentrations of HK *W. cibaria* CMU (HK-oraCMU) and CMS1 (HK-oraCMS1) (MOI = 1000) did not reduce cell viability, and the CFSs of *W. cibaria* CMU (CFS-oraCMU) and CMS1 (CFS-oraCMS1) did not reduce cell viability even at high concentrations (2 mg/mL) (Figure 1).

### 3.2. Effects of W. cibaria CMU and CMS1 on RANKL-Induced Osteoclastogenesis

#### 3.2.1. HK *W. cibaria* Inhibited Osteoclastogenesis and TRAP Activity in RAW 264.7 Cells

TRAP-positive, stained multinucleated cells were identified as osteoclasts (Figure 2). It was observed that RANKL induced RAW 264.7 cells to differentiate into mature multinucleated osteoclasts; however, the treatment with HK-oraCMU and HK-oraCMS1 significantly reduced osteoclast formation in a dose-dependent manner ($p < 0.05$) (Figure 2I). High concentrations of HK-oraCMU and HK-oraCMS1 (MOI = 1000) reduced osteoclast differentiation by 96.1% and 95.7%, respectively. TRAP activity was significantly increased by RANKL, and conversely, it was significantly decreased in the HK-oraCMU and HK-oraCMS1 treatment groups dose-dependently ($p < 0.05$) (Figure 2J). High concentrations of HK-oraCMU and HK-oraCMS1 (MOI = 1000) reduced TRAP activity by 46.0% and 50.9%, respectively.

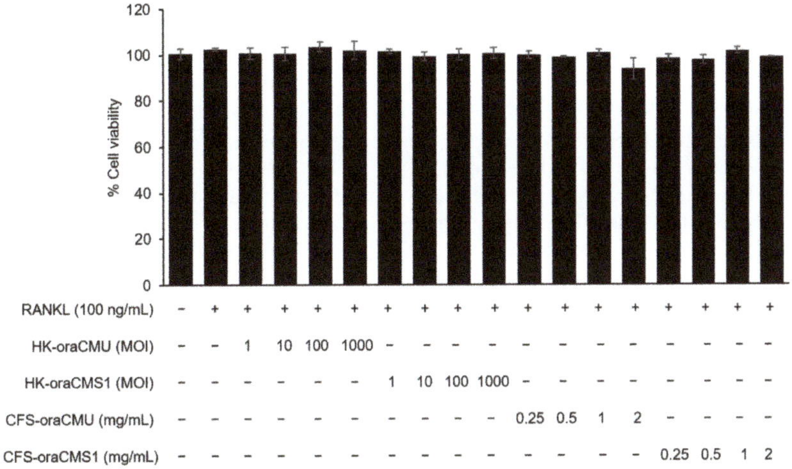


**Figure 1.** Effects of heat-killed *W. cibaria* CMU (HK-oraCMU), CMS1 (HK-oraCMS1), or the cell-free supernatants (CFSs) of *W. cibaria* CMU (CFS-oraCMU) and CMS1 (CFS-oraCMS1) on the viability of RAW 264.7 cells. All groups showed no statistically significant differences between groups ($p > 0.05$).

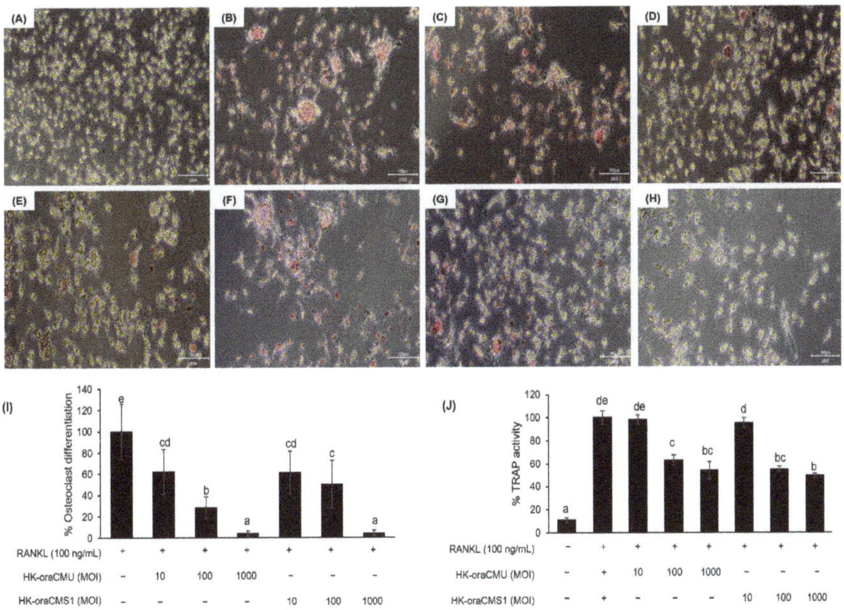


**Figure 2.** Dose-dependent inhibition of osteoclast differentiation by HK-oraCMU and HK-oraCMS1. RAW 264.7 cells were cultured in the presence of RANKL (100 ng/mL) and various concentrations of HK *W. cibaria* for 5 days. TRAP staining was observed using a microscope (scale bars = 100 μm; magnification: ×200). (**A–H**) Representative images of TRAP staining. (**A**) Untreated negative control;

(**B**) RANKL-treated control; (**C–E**) RANKL + HK-oraCMU (MOI = 10, 100, or 1000); (**F–H**) RANKL + HK-oraCMS1 (MOI = 10, 100, or 1000); (**I**) giant multinucleated cells containing ≥ 3 nuclei that stained positive for TRAP were identified as osteoclasts. Osteoclast differentiation was expressed as a percentage of the control (RANKL-treated only); (**J**) TRAP activity was determined at 540 nm. TRAP activity was expressed as a percentage of the control (RANKL-treated only). HK-oraCMU, heat-killed *W. cibaria* CMU; HK-oraCMS1, heat-killed *W. cibaria* CMS1. Different alphabet letters (a–e) indicate statistical differences as determined by ANOVA ($p < 0.05$).

### 3.2.2. CFSs of *W. cibaria* Inhibited Osteoclastogenesis and TRAP Activity in RAW 264.7 Cells

The effects of the CFS-oraCMU and CFS-oraCMS1 treatments on osteoclast differentiation are presented in Figure 3. Compared to the RANKL alone group, treatment with both CFS-oraCMU and CFS-oraCMS1 significantly inhibited differentiation into TRAP-positive multinucleated osteoclasts ($p < 0.05$) (Figure 3I). High concentrations of CFS-oraCMU and CFS-oraCMS1 (2 mg/mL) reduced osteoclast differentiation by 51.2% and 31.9%, respectively. Moreover, TRAP activity significantly decreased only in the high-concentration CFS-oraCMU treatment group (2 mg/mL), while all CFS-oraCMS1-treated groups exhibited decreased TRAP activity (Figure 3J). High concentrations of CFS-oraCMU and CFS-oraCMS1 reduced TRAP activity by 24.0% and 36.9%, respectively.

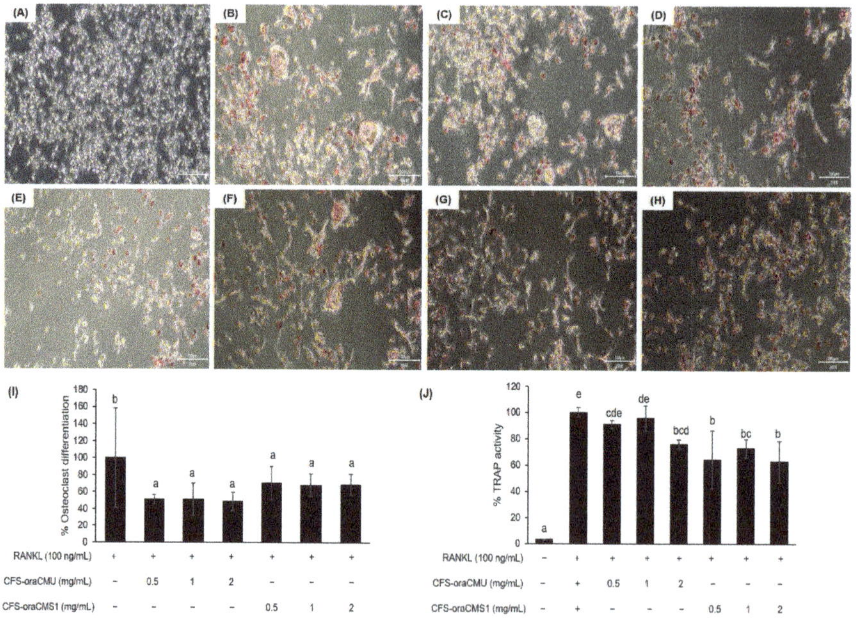


**Figure 3.** Inhibitory effects of the CFS-oraCMU and CFS-oraCMS1 on osteoclast differentiation. RAW 264.7 cells were incubated with RANKL (100 ng/mL) and various concentrations of CFS-oraCMU or CFS-oraCMS1 for 5 days. TRAP staining was observed using a microscope (scale bars = 100 μm; magnification: ×200). (**A–H**) Representative images of TRAP staining. (**A**) Untreated negative control; (**B**) RANKL-treated control; (**C–E**) RANKL + CFS-oraCMU (0.5, 1, or 2 mg/mL); (**F–H**) RANKL + CFS-oraCMS1 (0.5, 1, or 2 mg/mL); (**I**) giant multinucleated cells containing ≥ 3 nuclei that stained positive for TRAP were identified as osteoclasts. Osteoclast differentiation was expressed as a percentage of the control (RANKL-treated only); (**J**) TRAP activity was determined at 540 nm. TRAP activity was expressed as a percentage of the control (RANKL-treated only). CFS-oraCMU, cell-free supernatants of *W. cibaria* CMU; CFS-oraCSM1, cell-free supernatants of *W. cibaria* CMS1. Different alphabet letters (a–e) indicate statistical differences as determined by ANOVA ($p < 0.05$).

### 3.3. Effects of W. cibaria CMU and CMS1 on Bone Resorption

To determine whether bone resorption was inhibited by HK *W. cibaria*, pit formation was evaluated microscopically on dentin discs. The dentin section surface of the untreated negative control group appeared clean, whereas the group treated with RANKL alone exhibited large bone resorption pits (Figure 4B). Both HK-oraCMU and HK-oraCMS1 significantly inhibited pit formation by 99.0%, regardless of the concentration used (Figure 4O). Additionally, as shown in Figure 4I–N, pit formation was inhibited in both CFS-oraCMU and oraCMS1 groups compared to the RANKL-only group, and quantitative analysis confirmed that the pit formation was significantly suppressed (67.3–97.0%) in these groups compared to the RANKL-only control group ($p < 0.05$) (Figure 4O).

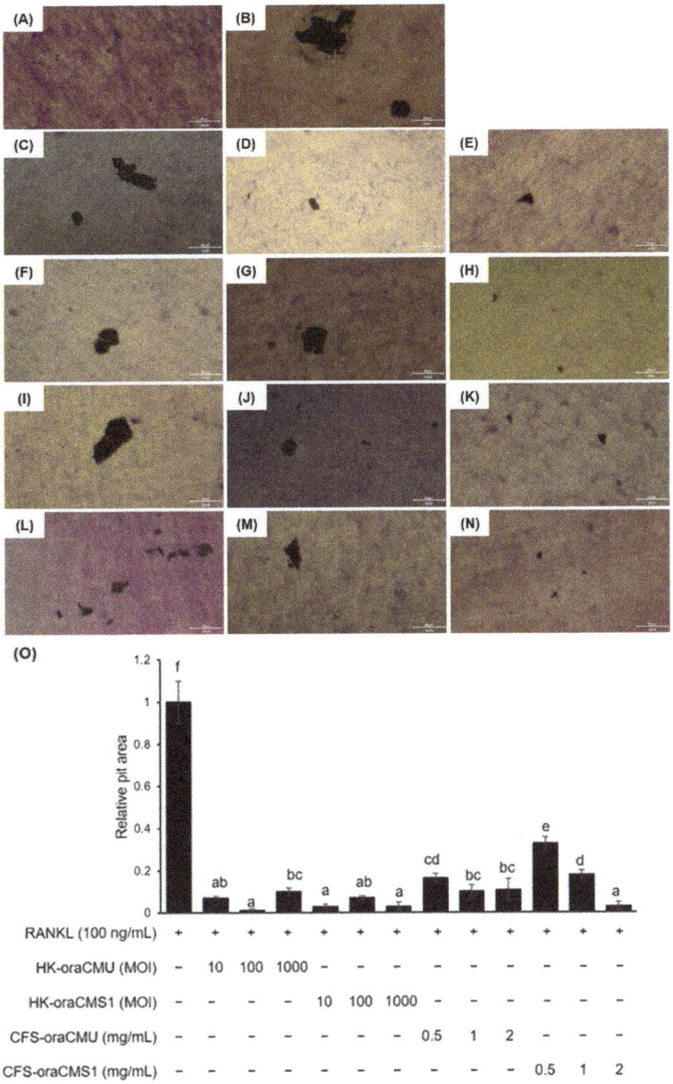


**Figure 4.** Inhibitory effects of *W. cibaria* CMU and CMS1 on bone resorption. (**A–N**) Representative images of the bone resorption pit. RAW 264.7 cells were incubated with RANKL (100 ng/mL) and

various concentrations of the test substances for 5 days. Bone resorption was observed using a microscope (scale bars = 50 μm; magnification: ×400). When the bone was resorbed, a pit was formed, which appeared black. (**A**) Untreated negative control; (**B**) RANKL-treated control; (**C–E**) RANKL + HK-oraCMU (MOI = 10, 100, or 1000); (**F–H**) RANKL + HK-oraCMS1 (MOI = 10, 100, or 1000); (**I–K**) RANKL + CFS-oraCMU (0.5, 1, or 2 mg/mL); (**L–N**) RANKL + CFS-oraCMS1 (0.5, 1, or 2 mg/mL); (**O**) quantitative analysis of bone resorption pit area. The pit area was expressed as a value relative to the RANKL-treated control. HK-oraCMU, heat-killed *W. cibaria* CMU; HK-oraCMS1, heat-killed *W. cibaria* CMS1; CFS-oraCMU, cell-free supernatants of *W. cibaria* CMU; CFS-oraCSM1, cell-free supernatants of *W. cibaria* CMS1. Different alphabet letters (a–f) indicate statistical differences as determined by ANOVA ($p < 0.05$).

### 3.4. *Effects of W. cibaria CMU and CMS1 on Osteoclastogenesis-Associated Gene Expression*

#### 3.4.1. HK *W. cibaria* Suppressed Osteoclastogenesis-Associated Gene Expression

We measured the mRNA expression of osteoclastogenesis-associated genes in RANKL-induced RAW 264.7 cells that were treated with various concentrations of HK *W. cibaria*. RANKL-induced in RAW 264.7 cells increased the mRNA expression of osteoclastogenesis-associated genes compared to that in the untreated negative control ($p < 0.05$) (Figure 5). Both HK-oraCMU and HK-oraCMS1 significantly inhibited the mRNA expression of c-Fos, NFATc1, OSCAR, DC-STAMP, cathepsin K, and TRAP in a dose-dependent manner compared to the RANKL-treated control (Figure 5A,B). In the HK-oraCMU group, the mRNA expression of c-Fos, NFATc1, OSCAR, DC-STAMP, cathepsin K, and TRAP was suppressed by 85.3%, 92.0%, 89.2%, 89.4%, 89.6%, and 96.8%, respectively. Additionally, HK-oraCMS1 suppressed the mRNA expression of c-Fos, NFATc1, OSCAR, DC-STAMP, cathepsin K, and TRAP by 88.7%, 85.3%, 97.5%, 80.9%, 90.3%, and 95.5%, respectively.

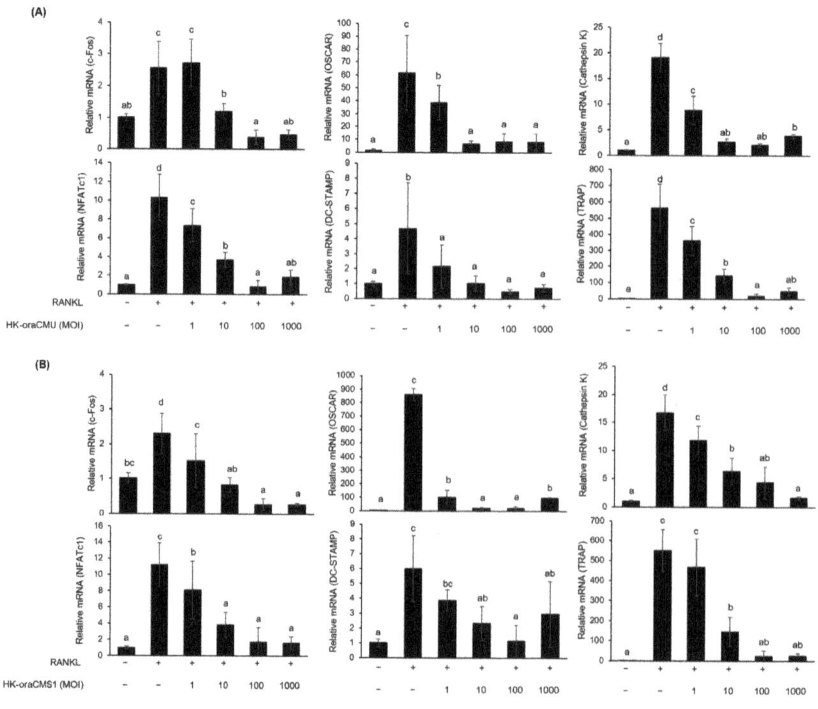


**Figure 5.** Inhibitory effects of HK-oraCMU (**A**) and HK-oraCMS1 (**B**) on the expression levels of osteoclast differentiation-associated genes in RANKL-stimulated RAW 264.7 cells. RAW 264.7 cells

were incubated with RANKL (100 ng/mL) and various concentrations of the test substances for 2 days. HK-oraCMU, heat-killed *W. cibaria* CMU; HK-oraCMS1, heat-killed *W. cibaria* CMS1. Relative gene expression was normalized to that of GAPDH. Different alphabet letters (a–d) indicate statistical differences as determined by ANOVA ($p < 0.05$).

#### 3.4.2. CFSs of *W. cibaria* Selectively Suppressed the Expression of Osteoclastogenesis-Associated Genes

The effects of CFS-oraCMU and CFS-oraCMS1 on the mRNA expression of osteoclas togenesis-associated genes are shown in Figure 6. CFS-oraCMU significantly suppressed the mRNA expression of NFATc1, OSCAR, DC-STAMP, and cathepsin K (Figure 6A). CFS-oraCMU inhibited the mRNA expression of NFATc1, OSCAR, DC-STAMP, and cathepsin K by 40.8%, 34.6%, 49.4%, and 57.5%, respectively. Similarly, CFS-oraCMS1 significantly downregulated the mRNA expression of c-Fos, NFATc1, OSCAR, and cathepsin K (Figure 6B). CFS-oraCMS1 suppressed the mRNA expression of c-Fos, NFATc1, OSCAR, and cathepsin K by 30.4%, 38.3%, 61.3%, and 63.5%, respectively.

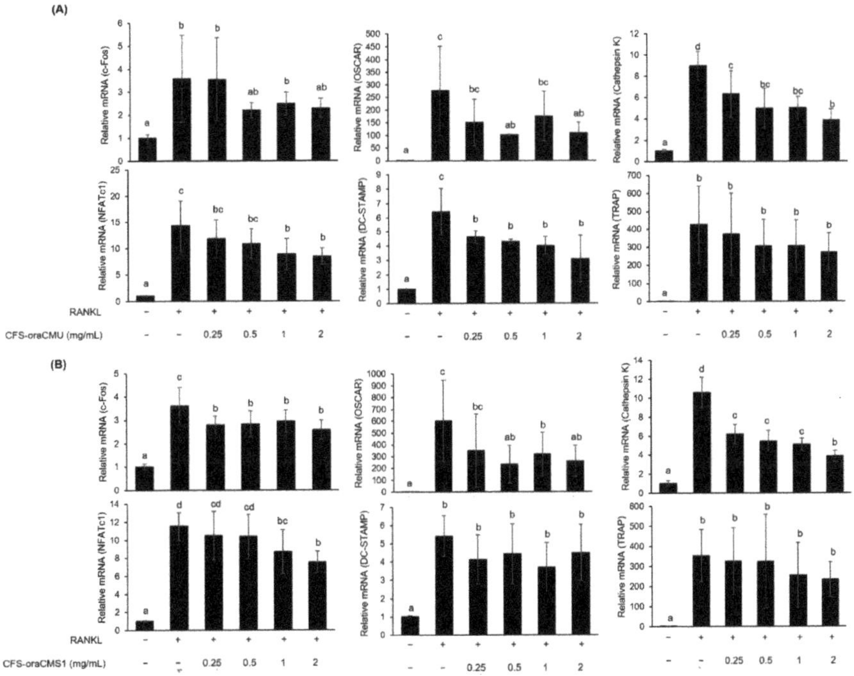


**Figure 6.** Inhibitory effects of CFS-oraCMU (**A**) and CFS-oraCMS1 (**B**) on the expression levels of osteoclast differentiation-associated genes in RNAKL-induced RAW 264.7 cells. RAW 264.7 cells were incubated with RANKL (100 ng/mL) and various concentrations of the test substances for 2 days. CFS-oraCMU, cell-free supernatants of *W. cibaria* CMU; CFS-oraCMS1, cell-free supernatants of *W. cibaria* CMS1. Relative gene expression was normalized to that of GAPDH. Different alphabet letters (a–d) indicate statistical differences as determined by ANOVA ($p < 0.05$).

### *3.5. Effects of W. cibaria CMU and CMS1 on Osteoclastogenesis-Associated Protein Expression*

#### 3.5.1. HK *W. cibaria* Suppressed Osteoclastogenesis-Associated Protein Expression

We subsequently examined the capacity of HK-oraCMU or HK-oraCMS1 to suppress the expression of osteoclastogenesis-associated proteins through Western blot analysis, with the results presented in Figure 7. RANKL (100 ng/mL) stimulation increased c-Fos, NFATc1, and cathepsin K protein expression in RAW264.7 cells, which was dose-dependently inhibited by treatment with HK-oraCMU or HK-oraCMS1. In contrast, protein expression was not decreased when RAW264.7 cells were cultured without direct contact with HK bacteria using a cell culture insert. When a high concentration of HK-oraCMU or HK-oraCMS1 (MOI = 100) was added onto the filter membrane of the cell culture insert, the expression levels of c-Fos, NFATc1, and cathepsin K were not significantly different from those of the positive control group treated with RANKL alone (Figure 7B). HK-oraCMU inhibited c-Fos, NFATc1, and Cathepsin K protein expression by 84.1%, 94.5%, and 69.5%, respectively, and HK-oraCMS1 inhibited the expression of these proteins by 77.5%, 91.9%, and 76.9%, respectively.

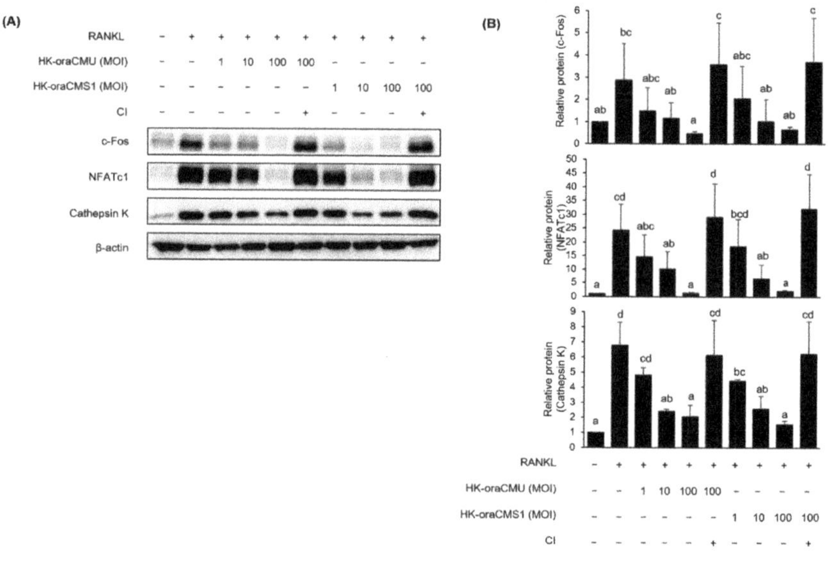


**Figure 7.** Inhibitory effects of HK-oraCMU and HK-oraCMS1 on the expression of osteoclast differentiation-associated proteins in RANKL-induced RAW 264.7 cells. RAW 264.7 cells were incubated for 2 days with RANKL (100 ng/mL) and various concentrations of the test substances. The protein expression levels were measured by Western blotting (**A**) and quantified (**B**). The quantification of the relative protein expression was normalized to the expression of β-actin. CI—cell culture inserts in the Transwell system. Different alphabet letters (a–d) indicate statistical differences as determined by ANOVA ($p < 0.05$).

3.5.2. CFSs of *W. cibaria* Suppressed Osteoclastogenesis-Associated Protein Expression

We further investigated the effect of CFS-oraCMU or CFS-oraCMU on the expression of osteoclastogenesis-associated proteins in RAW264.7 cells. Treatment with various concentrations of CFS decreased c-Fos, NFATc1, and cathepsin K protein expression (Figure 8). Moreover, it was confirmed that the protein expression level in the group treated with the highest concentration of CFS-oraCMS1 (2 mg/mL) was significantly lower than that in the positive control group treated with RANKL alone (Figure 8B). CFS-oraCMU inhibited c-Fos, NFATc1, and cathepsin K protein expression by 49.8%, 47.6%, and 33.1%, respectively, and CFS-oraCMS1 inhibited the expression of these proteins by 61.3%, 57.2%, and 46.8%, respectively.

### 3.6. *Effects of W. cibaria CMU and CMS1 on Cell Signaling Pathways*

Western blot analysis was performed to investigate whether HK *W. cibaria* or CFSs inhibits the activation of the NF-κB and MAPK pathways. RANKL (100 ng/mL) stimulation increased the phosphorylation of NF-κB and MAPK. Treatment with HK-oraCMU and HK-oraCMS1 for 5 min attenuated IκBα phosphorylation in both strains, and treatment for

15 min dose-dependently attenuated the phosphorylation of JNK and p38, but not ERK (Figure 9A). Following treatment with CFS-oraCMU and CFS-oraCMS1 for 5 min, IκBα phosphorylation was slightly weakened by both strains, and following 15 min of treatment, the phosphorylation of JNK and p38 was reduced in a dose-dependent manner. In addition, treatment with 2 mg/mL CFS-oraCMS1 attenuated ERK phosphorylation (Figure 9B).

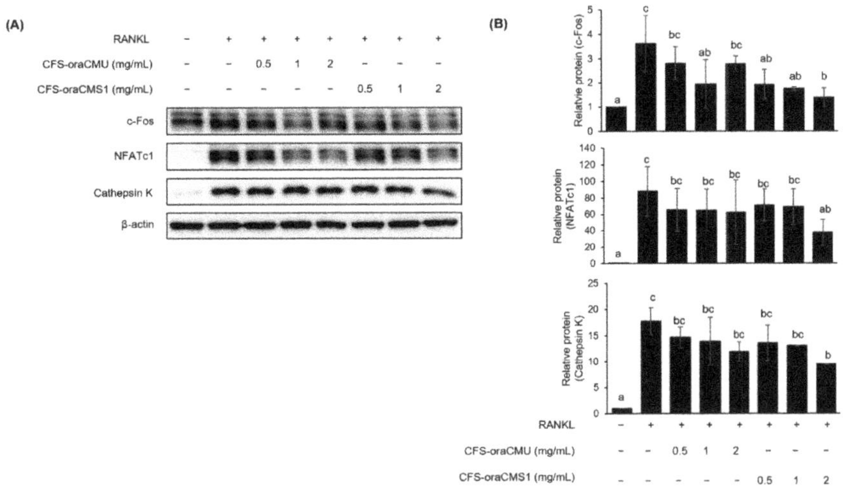


**Figure 8.** Inhibitory effects of CFS-oraCMU and CFS-oraCMS1 on the expression levels of osteoclast differentiation-associated proteins in RANKL-stimulated RAW 264.7 cells. RAW 264.7 cells were incubated with RANKL (100 ng/mL) and various concentrations of the test substances for 2 days. The protein expression levels were measured by Western blotting (**A**) and quantified (**B**). The protein expression was quantified and normalized to the expression of β-actin. Different alphabet letters (a–c) indicate statistical differences as determined by ANOVA ($p < 0.05$).

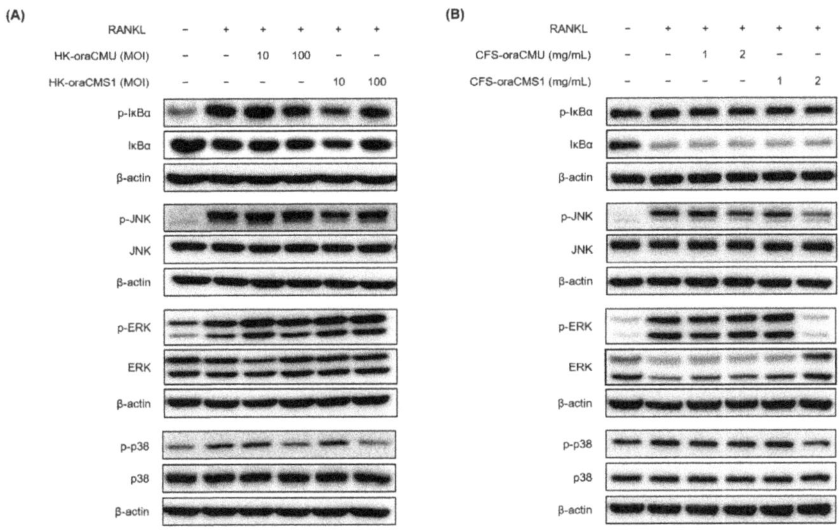


**Figure 9.** Inhibitory effects of heat-killed (**A**) or cell-free supernatants (**B**) of *W. cibaria* strains CMU and CMS1 on the expression levels of the NF-κB and MAPK signaling pathways in RANKL-stimulated

RAW 264.7 cells. RAW 264.7 cells were incubated with RANKL (100 ng/mL) and various concentrations of the test substances. Western blot analysis was used to measure the levels of proteins associated with NF-κB and MAPK activation, including IκBα, p38, JNK, and ERK, at 5 or 15 min. HK-oraCMU, heat-killed *W. cibaria* CMU; HK-oraCMS1, heat-killed *W. cibaria* CMS1; CFS-oraCMU, cell-free supernatants of *W. cibaria* CMU; CFS-oraCMS1, cell-free supernatants of *W. cibaria* CMS1. The relative protein expression was quantified and normalized to the expression of β-actin.

## 4. Discussion

Excessive osteoclast activity due to inflammation can lead to bone damage and loss, which can promote tissue damage and periodontal loss associated with periodontitis [3]. Therefore, if oral probiotics can inhibit the generation of osteoclasts, they are anticipated to suppress the occurrence and progression of periodontitis. Accordingly, several studies have evaluated probiotic therapeutic interventions as potential treatments to prevent alveolar bone loss.

These studies reported that probiotics inhibit bone loss in local inflammatory responses in conditions such as periodontitis, mostly using experimental animal models [16–18,26–28,31,32]. *Lactobacillus gasseri* and *Lactobacillus rhamnosus* were reported to reduce alveolar bone loss and inflammation scores in a periodontitis model induced by periodontal pathogens, including *P. gingivalis* [26,27]. In a ligation-induced rat periodontitis model, *Bifidobacterium animalis lactis*, *Bacillus subtilis*, and *Saccharomyces cerevisiae* were reported to reduce alveolar bone loss, reduce proinflammatory cytokines, lower osteoclast numbers, and alter the anaerobic–aerobic oral microbiota composition [31,32]. Additionally, *Lactobacillus brevis* and *W. cibaria* were reported to reduce alveolar bone loss, reduce proinflammatory cytokines, and reduce oral bacterial counts in a ligation-induced mouse periodontitis model [16,18].

Probiotics are defined as "live microorganisms that confer health benefits to the host when administered in adequate amounts" [33]. Although probiotics are being actively commercialized for their health benefits, technical limitations, such as viability control, place restrictions on production processes in various product development endeavors. Additionally, the safety profile of live strains when used in immunocompromised patients and pediatric populations has remained controversial [34,35].

Therefore, advancing research on functional foods in recent decades has given rise to a new conceptual term for microorganisms that have positive effects on their host. It has been reported that even non-viable microorganisms or the by-products of bacterial metabolism can exert biological activity in the host, leading to the emergence of new terms such as paraprobiotics and postbiotics [36]. Paraprobiotics are defined as "inactivated microbial cells that provide health benefits to the host" and include intact cell wall components with the cell structure intact. Postbiotics are defined as "probiotic metabolites that provide benefits to the host" and include CFSs containing proteins and organic acids secreted from viable cells [37].

CFSs from the postbiotics *Lactobacillus salivarius* and *Lactobacillus reuteri* have considerable potential as functional oral health ingredients to inhibit alveolar bone loss associated with periodontitis [25,38]. Meanwhile, the *W. cibaria* CMU (oraCMU) and CMS1 (oraCMS1) strains have been commercialized as oral probiotics, and there has been ongoing research on their beneficial effects on oral health [18–22]. In particular, oraCMU was confirmed to suppress alveolar bone loss in a periodontitis-induced animal model; however, research on other associated inhibitory mechanisms in addition to its anti-inflammatory mechanism has not yet been conducted [18]. Since there is a known correlation between osteoclasts and periodontitis, we sought to determine their inhibitory ability on osteoclastogenesis using HK bacteria as paraprobiotics as well as CFS-containing metabolites as postbiotics to gain a better understanding of the effect of oraCMU and oraCMS1 on osteoclastogenesis in vitro, in this study.

Osteoclasts are large, multinucleated cells that originate from hematopoietic stem cells [4–6]. Osteoclast differentiation is essentially regulated by RANKL [7,9]. RAW 264.7 cells are primarily used as osteoclast precursors because they differentiate into osteoclasts through RANKL stimulation [29,30]. RANK is present on the cell surface, and its expression is increased by RANKL, and differentiation into osteoclasts is promoted through the binding of RANK and RANKL. TRAP is abundantly distributed in the lysosomes of osteoclasts, and TRAP activity serves as a cytochemical marker that identifies multinucleated cells in bone tissue [39]. Therefore, in this study, TRAP staining was used to find out whether multinucleated osteoclasts were generated, and TRAP activity was measured as an indicator of the degree of osteoclast differentiation to investigate the effects of HK bacteria or metabolites.

We demonstrated that in the control group treated with RANKL alone, the formation of multinucleated osteoclasts was observed in RAW 264.7 cells, and treatment with both HK bacteria strains resulted in a significant dose-dependent reduction of both osteoclast formation and TRAP activity (Figure 2). Similarly, the CFSs also significantly reduced osteoclast formation and TRAP activity; however, even at a high concentration of 2 mg/mL, which did not exhibit cytotoxicity, its capacity to inhibit osteoclast formation was lower than that of HK bacteria (Figure 3). Therefore, it was confirmed that oraCMU and oraCMS1 suppress osteoclast differentiation by reducing TRAP activity, increased by RANKL, in both paraprobiotics and postbiotics.

Bone resorption activity is a typical characteristic of mature osteoclasts. Kyoi et al. [6] employed scanning electron microscopy to observe whether osteoclasts were formed on hard tissue sections to form microscopic pits on the surface. In our study, we evaluated pit formation using dentin discs to determine osteoclast function. The analysis revealed that the dentin disc surface of the untreated group was clean without pits, while the RANKL-only treated group exhibited larger pit sizes compared to the other treatment groups (Figure 4). These findings showed that osteoclasts can resorb bone and that the size of the bone pits increased proportionally with the number of osteoclasts. Analysis of the relative pit area revealed that it was significantly reduced in both the dead bacteria- and metabolite-treated groups. These findings suggest that both dead cells and metabolites effectively inhibit RANKL-mediated osteoclastogenesis and bone resorption.

Treating RAW 264.7 cells with RANKL initiates osteoclast differentiation signaling through RANK, thereby inducing the expression of key differentiation-related genes, including c-Fos, NFATc1, OSCAR, DC-STAMP, cathepsin K, and TRAP [7–13]. Furthermore, c-Fos and NFATc1 work synergistically to induce the expression of several key osteoclastogenesis-related genes [7–10]. Specifically, NFATc1 is known as a key transcriptional regulator of osteoclasts and is an essential factor for osteoclast differentiation [7–10]. OSCAR is an osteoclast-specific immune receptor that serves as a costimulatory signal essential for RANKL-mediated NFATc1 activation [11]. DC-STAMP is known to play a role in cell–cell fusion and is one of the fusion mediator molecules directly regulated by NFATc1 [13]. Meanwhile, cathepsin K and TRAP are cysteine proteases secreted by osteoclasts that are responsible for decomposing matrix proteins during bone resorption [12].

In this study, we found that HK-oraCMU and HK-oraCMS1 dose-dependently downregulated the gene expression levels of all analyzed osteoclast-specific genes, namely c-Fos, NFATc1, OSCAR, DC-STAMP, cathepsin K, and TRAP (Figure 5). Conversely, CFSs containing metabolites significantly suppressed gene expression levels for NFATc1, OSCAR, and cathepsin K but not TRAP, and this was observed for both strains (Figure 6). This is consistent with our findings that dead bacteria significantly downregulated all osteoclast-related genes, whereas metabolites only effectively downregulated certain genes. This result supports the fact that, when comparing TRAP results, dead bacteria exhibited a better inhibitory effect on osteoclast differentiation than metabolites. These results suggest that dead cells and their metabolites regulate osteoclastogenesis at different transcriptional levels.

Additionally, this study confirmed that dead bacteria significantly reduced the expression of all proteins dose-dependently, which is consistent with the results of gene expression inhibition (Figure 7). To determine the effect of direct interaction between HK bacteria and host cells, cell culture inserts with a pore size of 0.4 μm were used to prevent contact between RAW 264.7 cells and dead cells. Notably, HK bacteria did not inhibit the expression of proteins associated with osteoclast differentiation (Figure 7). These experiments confirmed that HK-oraCMU and HK-oraCMS1 require direct interaction with osteoclast precursors to inhibit osteoclastogenesis. These results are consistent with those of a previous study in which live *W. cibaria* CMU showed an anti-inflammatory effect on human gingival fibroblasts through direct contact [22]. Similarly, metabolites demonstrated a tendency to reduce the expression of proteins associated with osteoclast differentiation, with high concentrations of CFS-oraCMS1 inducing a significant decrease, which is consistent with the findings on gene expression analysis (Figure 8). Therefore, our results suggest that oraCMU and oraCMS1 inhibit osteoclastogenesis by downregulating the expression of proteins involved in osteoclast bone resorption and suppressing the expression of key downstream osteoclast differentiation-related genes, including NFATc1.

Osteoclast differentiation, mediated by RANKL/RANK binding, is achieved through a specific intracellular signaling pathway. Specifically, intracellular NF-κB and MAPK are activated, leading to increased expression of the transcription factor c-Fos, and subsequently, NFATc1, a crucial transcription factor for the entire osteoclast differentiation and formation process, is activated, resulting in osteoclast differentiation [4,9,10]. In this study, we confirmed through Western blotting analysis that HK bacteria and CFSs downregulate the phosphorylation of JNK and p38 during RANKL-induced osteoclast differentiation. These results suggest that oraCMU and oraCMS1 inhibit RANKL-induced osteoclastogenesis by blocking the MAPK signaling pathways, thereby downregulating the expression of the essential transcription factors c-Fos and NFATc1, subsequently downregulating various downstream-related genes including OSCAR, DC-STAMP, cathepsin K, and TRAP.

Taken together, our results demonstrated that paraprobiotic-like dead cells suppress osteoclastogenesis more effectively than postbiotic-like metabolites. Additionally, this study revealed that the dead bacterial strains CMU and CMS1 inhibited osteoclastogenesis through direct interaction with the host cells, which inactivated the MAPK signaling pathways, including JNK and p38, subsequently downregulating osteoclast-related genes. This suggests that periodontitis can be prevented by suppressing the expression of osteoclast-associated proteins, resulting in the suppression of osteoclast differentiation and bone resorption.

Paraprobiotics primarily reside in the bacterial cell envelope, encompassing a variety of molecules, including peptidoglycan, teichoic acid, cell wall polysaccharides, and cell surface-related proteins. These components are crucial effector molecules since they are the initial points of interaction with host cells [37]. Conversely, postbiotics refer to substances that bacteria secrete or release into the host environment following bacterial lysis, offering various physiological benefits to the host, including proteins, organic acids, and peptides [37]. However, this study has a few limitations. First, it was not possible to secure a sample size that could statistically determine the mechanism by which the HK bacteria and CFSs inactivate the NF-κB and MAPK pathways. Second, this study could not elucidate why CFS-oraCMS1 shows differentiation from other tested substances by downregulating ERK phosphorylation. Therefore, further research is needed in this study to specifically discuss strain-specific mechanisms of osteoclast formation inhibition in HK bacteria and CFSs.

## 5. Conclusions

The *W. cibaria* strains CMU and CMS1 can prevent periodontitis by suppressing osteoclast differentiation by inhibiting the expression of osteoclast-associated genes and proteins, thereby preventing bone destruction. Further research is needed to elucidate the effective in vitro components and modes of action of these bacteria against osteoclastogenesis.

**Author Contributions:** Conceptualization, M.-S.K.; methodology, G.-Y.P. and J.-A.P.; software, G.-Y.P.; validation, M.-S.K.; formal analysis, G.-Y.P. and M.-S.K.; investigation, G.-Y.P. and J.-A.P.; resources, M.-S.K.; data curation, G.-Y.P. and J.-A.P.; writing—original draft preparation, G.-Y.P., J.-A.P. and M.-S.K.; writing—review and editing, M.-S.K.; visualization, G.-Y.P.; supervision, M.-S.K.; project administration, M.-S.K.; funding acquisition, M.-S.K. All authors have read and agreed to the published version of the manuscript.

**Funding:** This research was funded by the Health Functional Food Development Support program (no. RS-2022-00167206), supported by the Ministry of Small and Medium Enterprises (SMEs) and Startups (MSS, Republic of Korea).

**Data Availability Statement:** The data presented in this study are available upon request from the corresponding author.

**Conflicts of Interest:** The authors are employees of OraTicx Inc.

## References

1. Nazir, M.A. Prevalence of periodontal disease, its association with systemic diseases and prevention. *Int. J. Health Sci.* **2017**, *11*, 72–80.
2. Kassebaum, N.J.; Bernabé, E.; Dahiya, M.; Bhandari, B.; Murray, C.J.; Marcenes, W. Global burden of severe periodontitis in 1990–2010: A systematic review and meta-regression. *J. Dent. Res.* **2014**, *93*, 1045–1053. [CrossRef] [PubMed]
3. Li, Y.; Ling, J.; Jiang, Q. Inflammasomes in alveolar bone loss. *Front. Immunol.* **2021**, *12*, 691013. [CrossRef]
4. Takayanagi, H. Osteoimmunology: Shared mechanisms and crosstalk between the immune and bone systems. *Nat. Rev. Immunol.* **2007**, *7*, 292–304. [CrossRef]
5. Datta, H.K.; Ng, W.F.; Walker, J.A.; Tuck, S.P.; Varanasi, S.S. The cell biology of bone metabolism. *J. Clin. Pathol.* **2008**, *61*, 577–587. [CrossRef] [PubMed]
6. Kiyoi, T. Bone resorption activity in mature osteoclasts. *Methods Mol. Biol.* **2018**, *1868*, 215–222. [CrossRef] [PubMed]
7. Wada, T.; Nakashima, T.; Hiroshi, N.; Penninger, J.M. RANKL-RANK signaling in osteoclastogenesis and bone disease. *Trends Mol. Med.* **2006**, *12*, 17–25. [CrossRef] [PubMed]
8. Lu, S.H.; Chen, T.H.; Chou, T.C. Magnolol inhibits RANKL-induced osteoclast differentiation of raw 264.7 macrophages through heme oxygenase-1-dependent inhibition of NFATc1 expression. *J. Nat. Prod.* **2015**, *78*, 61–68. [CrossRef]
9. Ding, D.; Yan, J.; Feng, G.; Zhou, Y.; Ma, L.; Jin, Q. Dihydroartemisinin attenuates osteoclast formation and bone resorption via inhibiting the NF-κB, MAPK and NFATc1 signaling pathways and alleviates osteoarthritis. *Int. J. Mol. Med.* **2022**, *49*, 4. [CrossRef]
10. Hussein, H.; Boyaka, P.; Dulin, J.; Russell, D.; Smanik, L.; Azab, M.; Bertone, A.L. Cathepsin K localizes to equine bone in vivo and inhibits bone marrow stem and progenitor cells differentiation in vitro. *J. Stem Cells Regen. Med.* **2017**, *13*, 45–53. [CrossRef]
11. Gu, H.; An, H.J.; Kim, J.Y.; Kim, W.H.; Gwon, M.G.; Kim, H.J.; Han, S.M.; Park, I.; Park, S.C.; Leem, J.; et al. Bee venom attenuates *Porphyromonas gingivalis* and RANKL-induced bone resorption with osteoclastogenic differentiation. *Food Chem. Toxicol.* **2019**, *129*, 344–353. [CrossRef] [PubMed]
12. Cheng, Y.; Liu, H.; Li, J.; Ma, Y.; Song, C.; Wang, Y.; Li, P.; Chen, Y.; Zhang, Z. Evaluation of culture conditions for osteoclastogenesis in RAW264.7 cells. *PLoS ONE* **2022**, *17*, e0277871. [CrossRef] [PubMed]
13. Lee, A.S.; Sung, M.J.; Son, S.J.; Han, A.R.; Hong, S.M.; Lee, S.H. Effect of menaquinone-4 on receptor activator of nuclear factor κB ligand-induced osteoclast differentiation and ovariectomy-induced bone loss. *J. Med. Food* **2023**, *26*, 128–134. [CrossRef]
14. AlQranei, M.S.; Chellaiah, M.A. Osteoclastogenesis in periodontal diseases: Possible mediators and mechanisms. *J. Oral Biosci.* **2020**, *62*, 123–130. [CrossRef] [PubMed]
15. Hathaway-Schrader, J.D.; Novince, C.M. Maintaining homeostatic control of periodontal bone tissue. *Periodontology 2000* **2021**, *86*, 157–187. [CrossRef] [PubMed]
16. Maekawa, T.; Hajishengallis, G. Topical treatment with probiotic *Lactobacillus brevis* CD2 inhibits experimental periodontal inflammation and bone loss. *J. Periodontal Res.* **2014**, *49*, 785–791. [CrossRef] [PubMed]
17. Oliveira, L.F.; Salvador, S.L.; Silva, P.H.; Furlaneto, F.A.; Figueiredo, L.; Casarin, R.; Ervolino, E.; Palioto, D.B.; Souza, S.L.; Taba, M.J.; et al. Benefits of *Bifidobacterium animalis* subsp. *lactis* probiotic in experimental periodontitis. *J. Periodontol.* **2017**, *88*, 197–208. [CrossRef]
18. Kim, J.W.; Jung, B.H.; Lee, J.H.; Yoo, K.Y.; Lee, H.; Kang, M.S.; Lee, J.K. Effect of *Weissella cibaria* on the reduction of periodontal tissue destruction in mice. *J. Periodontol.* **2020**, *91*, 1367–1374. [CrossRef]
19. Kang, M.S.; Yeu, J.E.; Hong, S.P. Safety Evaluation of oral care probiotics *Weissella cibaria* CMU and CMS1 by phenotypic and genotypic analysis. *Int. J. Mol. Sci.* **2019**, *20*, 2693. [CrossRef]
20. Kim, M.J.; You, Y.O.; Kang, J.Y.; Kim, H.J.; Kang, M.S. *Weissella cibaria* CMU exerts an anti-inflammatory effect by inhibiting *Aggregatibacter actinomycetemcomitans*-induced NF-κB activation in macrophages. *Mol. Med. Rep.* **2020**, *22*, 4143–4150. [CrossRef]
21. Kang, M.S.; Lee, D.S.; Lee, S.A.; Kim, M.S.; Nam, S.H. Effects of probiotic bacterium *Weissella cibaria* CMU on periodontal health and microbiota: A randomised, double-blind, placebo-controlled trial. *BMC Oral Health* **2020**, *20*, 243. [CrossRef] [PubMed]
22. Kang, M.S.; Park, G.Y.; Lee, A.R. In vitro preventive effect and mechanism of action of *Weissella cibaria* CMU against *Streptococcus mutans* biofilm formation and periodontal pathogens. *Microorganisms* **2023**, *11*, 962. [CrossRef] [PubMed]
23. Liu, T.H.; Tsai, T.Y.; Pan, T.M. The anti-periodontitis effects of ethanol extract prepared using *Lactobacillus paracasei* subsp. *paracasei* NTU 101. *Nutrients* **2018**, *10*, 472. [CrossRef] [PubMed]

24. Quach, D.; Parameswaran, N.; McCabe, L.; Britton, R.A. Characterizing how probiotic *Lactobacillus reuteri* 6475 and lactobacillic acid mediate suppression of osteoclast differentiation. *Bone Rep.* **2019**, *11*, 100227. [CrossRef]
25. Jung, J.I.; Baek, S.M.; Nguyen, T.H.; Kim, J.W.; Kang, C.H.; Kim, S.; Imm, J.Y. Effects of probiotic culture supernatant on cariogenic biofilm formation and RANKL-induced osteoclastogenesis in RAW 264.7 macrophages. *Molecules* **2021**, *26*, 733. [CrossRef] [PubMed]
26. Kobayashi, R.; Kobayashi, T.; Sakai, F.; Hosoya, T.; Yamamoto, M.; Kurita-Ochiai, T. Oral administration of *Lactobacillus gasseri* SBT2055 is effective in preventing *Porphyromonas gingivalis*-accelerated periodontal disease. *Sci. Rep.* **2017**, *7*, 545. [CrossRef] [PubMed]
27. Gatej, S.M.; Marino, V.; Bright, R.; Fitzsimmons, T.R.; Gully, N.; Zilm, P.; Gibson, R.J.; Edwards, S.; Bartold, P.M. Probiotic *Lactobacillus rhamnosus* GG prevents alveolar bone loss in a mouse model of experimental periodontitis. *J. Clin. Periodontol.* **2018**, *45*, 204–212. [CrossRef]
28. Ricoldi, M.S.T.; Furlaneto, F.A.C.; Oliveira, L.F.F.; Teixeira, G.C.; Pischiotini, J.P.; Moreira, A.L.G.; Ervolino, E.; de Oliveira, M.N.; Bogsan, C.S.B.; Salvador, S.L.; et al. Effects of the probiotic *Bifidobacterium animalis* subsp. *lactis* on the non-surgical treatment of periodontitis. A histomorphometric, microtomographic and immunohistochemical study in rats. *PLoS ONE* **2017**, *12*, e0179946. [CrossRef]
29. Kamio, N.; Kawato, T.; Tanabe, N.; Kitami, S.; Morita, T.; Ochiai, K.; Maeno, M. Vaspin attenuates RANKL-induced osteoclast formation in RAW264.7 cells. *Connect. Tissue Res.* **2013**, *54*, 147–152. [CrossRef]
30. Song, C.; Yang, X.; Lei, Y.; Zhang, Z.; Smith, W.; Yan, J.; Kong, L. Evaluation of efficacy on RANKL induced osteoclast from RAW264.7 cells. *J. Cell. Physiol.* **2019**, *234*, 11969–11975. [CrossRef]
31. Garcia, V.G.; Knoll, L.R.; Longo, M.; Novaes, V.C.; Assem, N.Z.; Ervolino, E.; de Toledo, B.E.; Theodoro, L.H. Effect of the probiotic *Saccharomyces cerevisiae* on ligature-induced periodontitis in rats. *J. Periodontal Res.* **2016**, *51*, 26–37. [CrossRef] [PubMed]
32. Messora, M.R.; Pereira, L.J.; Foureaux, R.; Oliveira, L.F.; Sordi, C.G.; Alves, A.J.; Napimoga, M.H.; Nagata, M.J.; Ervolino, E.; Furlaneto, F.A. Favourable effects of *Bacillus subtilis* and *Bacillus licheniformis* on experimental periodontitis in rats. *Arch. Oral Biol.* **2016**, *66*, 108–119. [CrossRef]
33. Reid, G. Food and Agricultural Organization of the United Nation and the WHO. The importance of guidelines in the development and application of probiotics. *Curr. Pharm. Des.* **2005**, *11*, 11–16. [CrossRef] [PubMed]
34. Boyle, R.J.; Robins-Browne, R.M.; Tang, M.L. Probiotic use in clinical practice: What are the risks? *Am. J. Clin. Nutr.* **2006**, *83*, 1256–1264, quiz 1446–1447. [CrossRef]
35. Deshpande, G.; Athalye-Jape, G.; Patole, S. Para-probiotics for preterm neonates-The next frontier. *Nutrients* **2018**, *10*, 871. [CrossRef] [PubMed]
36. Piqué, N.; Berlanga, M.; Miñana-Galbis, D. Health benefits of heat-killed (tyndallized) probiotics: An overview. *Int. J. Mol. Sci.* **2019**, *20*, 2534. [CrossRef]
37. Teame, T.; Wang, A.; Xie, M.; Zhang, Z.; Yang, Y.; Ding, Q.; Gao, C.; Olsen, R.E.; Ran, C.; Zhou, Z. Paraprobiotics and postbiotics of probiotic *Lactobacilli*, their positive effects on the host and action mechanisms: A review. *Front. Nutr.* **2020**, *7*, 570344. [CrossRef]
38. Jeong, Y.J.; Jung, J.I.; Kim, Y.; Kang, C.H.; Imm, J.Y. Effects of *Lactobacillus reuteri* MG5346 on receptor activator of nuclear factor-kappa B ligand (RANKL)-induced osteoclastogenesis and ligature-induced experimental periodontitis rats. *Food Sci. Anim. Resour.* **2023**, *43*, 157–169. [CrossRef]
39. Ballanti, P.; Minisola, S.; Pacitti, M.T.; Scarnecchia, L.; Rosso, R.; Mazzuoli, G.F.; Bonucci, E. Tartrate-resistant acid phosphate activity as osteoclastic marker: Sensitivity of cytochemical assessment and serum assay in comparison with standardized osteoclast histomorphometry. *Osteoporos. Int.* **1997**, *7*, 39–43. [CrossRef]

Journal of
*Functional*
*Biomaterials*

MDPI

*Article*

# Osteoinductive Properties of Autologous Dentin: An Ex Vivo Study on Extracted Teeth

**Giulia Mazzucchi [1], Alessia Mariano [2], Giorgio Serafini [1], Luca Lamazza [1], Anna Scotto d'Abusco [2], Alberto De Biase [1,*] and Marco Lollobrigida [1]**

1 Department of Oral and Maxillofacial Sciences, Sapienza University of Rome, Via Caserta 6, 00161 Rome, Italy; giulia.mazzucchi@uniroma1.it (G.M.); giorgio.serafini@uniroma1.it (G.S.); luca.lamazza@uniroma1.it (L.L.); marco.lollobrigida@uniroma1.it (M.L.)

2 Department of Biochemical Sciences "Alessandro Rossi Fanelli", Sapienza University of Rome, Piazzale Aldo Moro 5, 00185 Rome, Italy; alessia.mariano@uniroma1.it (A.M.); anna.scottodabusco@uniroma1.it (A.S.d.)

* Correspondence: alberto.debiase@uniroma1.it; Tel.: +39-06-49976626

**Abstract:** Over the last decades, a variety of biomaterials, ranging from synthetic products to autologous and heterologous grafts, have been recommended to conserve and regenerate bone tissue after tooth extraction. We conducted a biochemical study on ground extracted teeth that aimed to evaluate the osteoinductive and osteoconductive potential of dentin by assessing the releases of bone morphogenetic protein (BMP-2), osteocalcin (OC) and osteonectin (ON) over time (24 h, 10 days and 28 days). Twenty-six patients, who required the extraction of nonrestorable teeth, were enrolled in the study according to the inclusion criteria, as follows: thirteen young patients 18 to 49 years of age (UNDER 50), and thirteen patients of 50 to 70 years (OVER 50); a total of twenty-six teeth were extracted, ground and analyzed by enzyme-linked immunosorbent assays (ELISA). All ground teeth released BMP-2, OC and ON at each time point; no differences were observed between the UNDER-50 and OVER-50 patients. The results of the study support the use of autologous dentin as osteoinductive material for bone regeneration procedures, irrespective of patients' ages.

**Keywords:** biomaterials; osteoinductive properties; autologous dentin graft; Smart Dentin Grinder®

## 1. Introduction

Over the past few decades, regenerative medicine has gained significant attention in the dental field. Guided bone regeneration (GBR) in dentistry traces its roots to the 1980s, evolving from principles used in periodontics [1]. Initially, the focus was on using barrier membranes to facilitate bone growth by preventing soft tissue invasion into bone defects. However, a significant milestone in the advancement of GBR was the development and use of bone substitutes, which have dramatically enhanced the effectiveness and applications of this technique. Bone substitutes were introduced as a solution to the limitations of autogenous bone grafts, which required harvesting bone from a patient's own body, often resulting in additional surgery and morbidity. The early bone substitutes included allografts, derived from donor bone, and xenografts, sourced from other species, such as bovine bone [2–6]. These materials provided a scaffold for new bone growth, but their success is limited by factors like the inert nature and varying resorption rates. To overcome such limitations, the integration of growth factors and biologics into bone substitutes has been proposed [7]. By incorporating elements such as bone morphogenetic proteins (BMPs) and other biomolecules, these bone substitutes not only serve as scaffolds but also actively promote and accelerate the bone healing process [8]. However, biofunctionalizing bone substitutes is costly and challenging to apply in most clinical situations.

Graft materials should have several ideal properties; they should be biocompatible, resorbable, osteoconductive, osteoinductive, easy to use and cost effective [9]. Autologous bone presents the majority of these characteristics [10–15] but needs a donor site, resulting in increased morbidity. Moreover, the reduced or null cell viability due to the harvesting procedure and, in some cases, the high resorption rate, discourages the use of autologous bone in most cases [16]. Other biomaterials with different rates of reabsorption have been suggested but, in most cases, such materials have only osteoconductive properties [7,17].

In recent years, dentine has been proposed as an alternative graft material for bone regeneration to conventional bone substitutes and autologous bone [18–21]. The use of dentin as an autologous graft in the form of a demineralized dentin matrix (DDM) has emerged because its biological composition closely resembles that of alveolar bone [22–24]. Several similarities exist between teeth and bone, including embryological origin, both deriving from neural crest cells, and structural similarities, though with some differences [25–28]. Alveolar bone comprises approximately 65% inorganic and 25% organic materials. Conversely, dentin consists of 70–75% inorganic and 20% organic components. In addition, both dentin and bone contain type I collagen (90%), and noncollagenous proteins (10%) consisting of osteonectin (ON), dentin sialoprotein (DSP), bone sialoprotein (BSP), osteocalcin (OC), and bone morphogenetic protein (BMP), which are essential for the formation and mineralization of the bone matrix [29–33].

Osteocalcin is produced by osteoblasts and is involved in several biologic processes, including the regulation of bone matrix mineralization and the formation of hydroxyapatite crystals. It also plays a central role in cell adhesion and signaling, facilitating the adhesion of osteoblasts to the bone matrix, promoting the formation and repair of dental tissue and acting as a growth and differentiation factor for osteoblasts [34,35].

Osteonectin is another noncollagenous protein of the bone matrix, which is produced by osteoblasts and odontoblasts. Osteonectin is contained in both dentin and enamel, with a higher concentration in dentin. It binds to collagen, providing strength and flexibility to the dental tissue. Osteonectin also plays a role in the mineralization of dentin and enamel, promoting the formation of hydroxyapatite crystals [34,35].

BMPs have also been identified in the human dentin matrix after demineralization [33]. BMPs have osteoinductive capacity, promoting the differentiation of mesenchymal stem cells into osteoblasts [36]. Although bone-derived BMPs and dentin-derived BMPs have different biochemical structures, they have similar functions in regulating the development and repair of osseous and dental tissues [23,32,33].

DDM primarily functions as a scaffold to support cell migration (osteoconduction) but may also elicit an osteoinductive action due to the presence of dentinal BMP-2 and other molecules. It has been observed, however, that mineral components of dentin could trap the BMPs limiting their bioavailability. Demineralization has been suggested as a solution for the release of a higher amount of growth factors and noncollagenous proteins [37]. Both demineralized bone matrix (DBM) and demineralized dentin matrix (DDM) retain type I collagen, growth factors and BMP-2 [18,36].

Considering the abovementioned biological properties of DDM, extracted teeth were first utilized as autologous graft materials for regenerative procedures [38]. Autologous dentine can be inserted into the postextractive or atrophic sites of the same patient without causing inflammation or rejection and is gradually resorbed and replaced by new bone [39,40].

To further support the use of ground teeth as osteoinductive graft material, the aim of this study was to investigate the release over time of BMP-2, OC and ON from ground extracted teeth.

## 2. Materials and Methods

### 2.1. Patient Recruitment, Inclusion and Exclusion Criteria

Twenty-six patients who required the extraction of nonrestorable teeth were included in the study, as follows: 13 patients aged between 18 and 49 years (UNDER 50 group), and 13 patients aged between 50 and 70 years (OVER 50 group).

The donor patients were enrolled according to the following inclusion criteria:

- Patients in need of dental extraction due to severe caries, coronal and/or radicular fracture, nontreatable periapical lesion, grade 2 or higher mobility, untreatable periodontal defects;
- Patients in need of extraction of impacted mandibular third molars with partial/total bony/osteo-mucosal inclusion, following recurrent abscesses, pericoronitis, severe caries affecting the third molar, severe caries affecting the adjacent second molar that cannot be otherwise treated without prior extraction of the third molar, distal periodontal defects of the adjacent second molar, root resorption affecting the adjacent second molar, dysplastic lesions affecting the mandibular third molar;
- Patients willing and able to provide written informed consent;
- Nonsmoker or light smoker (less than 10 cigarettes per day);
- Absence of contraindications to materials and anesthesia;
- Good overall health.

Patients who required extraction of teeth with root filling material were excluded from the study.

### 2.2. Ethics Statement

The study followed the principles of the Declaration of Helsinki in relation to the research work involving human species, and it received approval from the Ethics Committee of Policlinico Umberto I of Rome, Italy (protocol code: 5456/2019, 25 July 2019). Patients were enrolled in the study after signing a written informed consent form.

### 2.3. Grinding of Teeth

Immediately after the extraction, teeth were removed of enamel, cement, remaining calculus, caries, and periodontal ligament using a tungsten carbide bur and then dried with an air syringe. Obtained teeth were then ground for 3 s and sorted for 10 s using a dedicated device (Smart Dentin Grinder™, KometaBio Inc., Cresskill, NJ, USA). After grinding, the granules produced were sorted into the following two compartments with drawers: an upper compartment and a lower compartment. Granules between 300 and 1200 μm in diameter with a porosity of 2–14 μm accumulated in the upper compartment, while granules smaller than 300 μm accumulated in the lower compartment [41,42]. A 0.5 g sample of granules was taken from the upper drawer compartment and placed in a sterile Dappen dish, soaked with 0.5 M NaOH and 20% ethanol solution for 10 min to dissolve organic remains from the dentinal tubules and rinsed two times (3 min each) with phosphate-buffered saline solution (PBS). The granules were then immersed in 2 mL PBS solution and stored at 4 °C. The conditioned PBS was then collected and analyzed at 24 h (T1), 10 days (T2) and 28 days (T3) for the quantification of the BMP-2, osteocalcin (OC) and osteonectin (ON).

### 2.4. ELISA Tests

The amount of BMP-2, OC and ON released in the PBS was determined using enzyme-linked immunosorbent assay kits (Fine Test ELISA, Fine Biotech Co., Ltd., Wuhan, China), according to the manufacturer's instructions. Briefly, 100 μL of conditioned PBS was added to the wells and analyzed in duplicate. The plate was static incubated for 90 min at 37 °C. Then, the plate was washed twice, and 100 μL of biotin-labeled antibody was added to each well and incubated for 60 min at 37 °C. After incubation, the plate was washed three times and incubated with 100 μL of HRP–streptavidin conjugate for 30 min at 37 °C. Then, the plate was washed five times and 90 μL of tetramethylbenzidine substrate was added and incubated for 15 min at 37 °C to develop the blue color. Finally, 50 μL of an acid solution was added to turn the color to yellow. The optical density (O.D.) absorbance was measured at 450 nm by a microplate reader (NB-12-0035, NeoBiotech, Holden, MA, USA).

### 2.5. Statistical Analysis

The statistical analysis was performed using a nonparametric one-way ANOVA (Kruskal–Wallis test) and post hoc Mann–Whitney *U* (MWU) test (adjusted by Bonferroni correction for multiple tests) and applied on the original data to obtain a comparison among groups at each time point. The data were plotted in histograms using Prism 5.0 software (GraphPad Software, San Diego, CA, USA).

## 3. Results and Discussion

In order to determine the odonto/osteoinductive capacity of the ground teeth, the release of three factors (i.e., BMP-2, OC and ON), which play important roles in matrix organization, mineralization and bone formation, was evaluated. The amount of BMP-2, OC and ON released in PBS was measured at 24 h (T1), 10 days (T2) and 28 days (T3) after extraction. As shown in Figures 1–3, all growth factors had already been released after 24 h and continued until day 28. An interindividual difference in the amount of factor release was observed without statistically significant differences.

The patients in the UNDER 50 group showed a release of BMP-2 $\leq$ 200 pg/mL, at three time points; only two patients showed a release of about 300 pg/mL (Figure 1A). All patients in the OVER 50 group showed a release $\leq$ 200 pg/mL (Figure 1B). In most samples, the release was constant over time; only one patient showed an increasing release from T1 to T3 (Figure 1B).

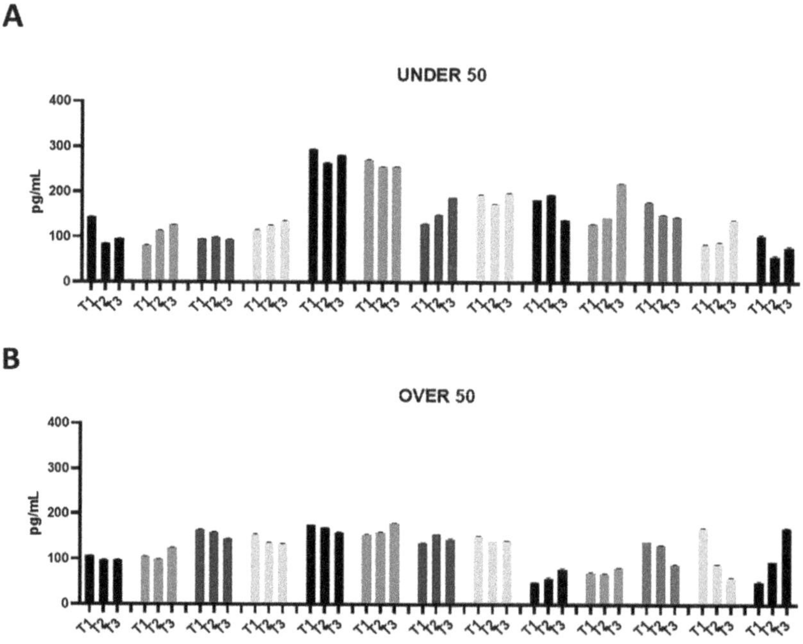


**Figure 1.** Bone morphogenetic protein (BMP)-2 releases over time in the UNDER 50 and OVER 50 patients. The BMP-2 release in the ground teeth was determined at T1 (24 h), T2 (10 days) and T3 (28 days) by an ELISA test: (**A**) UNDER 50 group with 13 patients; (**B**) OVER 50 group with 13 patients. The data from each patient are grouped and reported as different grey shades (T1, T2 and T3). The results are expressed as the mean ± standard deviation of the data obtained by duplicate analyses, performed by Prism 5.0 software (GraphPad).

All of the analyzed ground teeth had higher productions of OC compared to BMP-2. The BMP-2 was released in a concentration of picograms, whereas the OC was released in a concentration of nanograms. Most specifically, the OVER 50 group showed comparable releases of around 10 ng/mL, while in the UNDER 50 group a more interindividual variability is reported (Figure 2A,B). In the UNDER 50 group, four patients showed a very low release of OC, approximately 0.5 ng/mL, at three time points (Figure 2A). In the OVER 50 group, all samples showed a constant release at all analyzed times, except for three samples that exhibited decreases over time, in particular after 28 days, with a final concentration lower than 0.5 ng/mL (Figure 2B).

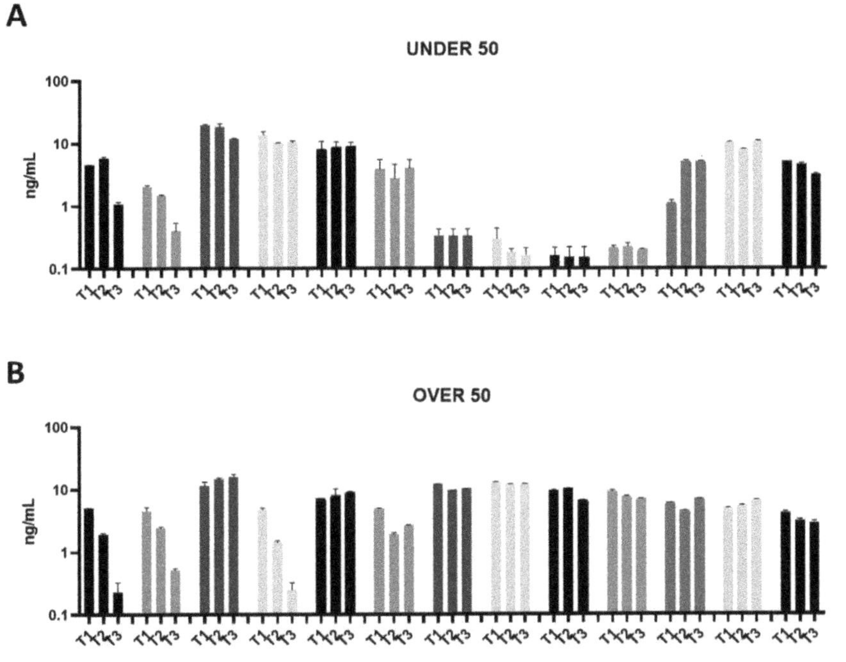


**Figure 2.** Osteocalcin (OC) releases over time in the UNDER 50 and OVER 50 patients. The OC release in the ground teeth was determined at T1 (24 h), T2 (10 days) and T3 (28 days) by an ELISA test: (**A**) UNDER 50 group with 13 patients; (**B**) OVER 50 group with 13 patients. The data from each patient are grouped and reported as different grey shades (T1, T2 and T3). The results are expressed as mean ± Standard Deviation of data obtained by duplicate analyses, performed by Prism 5.0 software (GraphPad).

The ON release was assessed in a range of approximately 0.2–0.6 ng/mL in all the samples from both groups (Figure 3A,B). Moreover, the release remained constant over time. Compared to OC, ON did not show an interindividual variability.

To further verify whether statistical differences between the UNDER 50 and OVER 50 group could be observed, a comparative analysis was performed. The medium releases of the three growth factors for each analyzed time are reported in Figure 4, with light grey representing the UNDER 50 group and dark grey the OVER 50 group. BMP-2 and ON did not show any difference between the groups and over time. The release of the OC factor was higher in the OVER 50 patients compared to the UNDER 50 subjects, for all analyzed times. The main difference was observed for T1, even if the differences were not statistically significant (T1, $p = 0.06$; T2, $p = 0.22$; T3, $p = 0.36$).

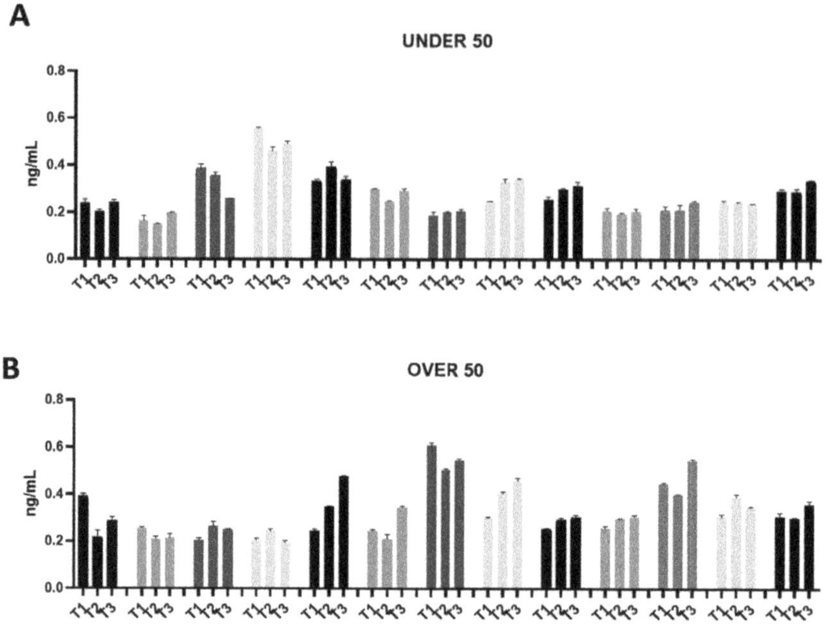


**Figure 3.** Osteonectin (ON) releases over time in the UNDER 50 and OVER 50 patients. The ON release in the ground teeth was determined at T1 (24 h), T2 (10 days) and T3 (28 days) by an ELISA test: (**A**) UNDER 50 group with 13 patients; (**B**) OVER 50 group with 13 patients. The data from each patient are grouped and reported as different shades of grey (T1, T2 and T3). The results are expressed as the mean ± standard deviation of data obtained by duplicate analyses, performed using Prism 5.0 software (GraphPad).

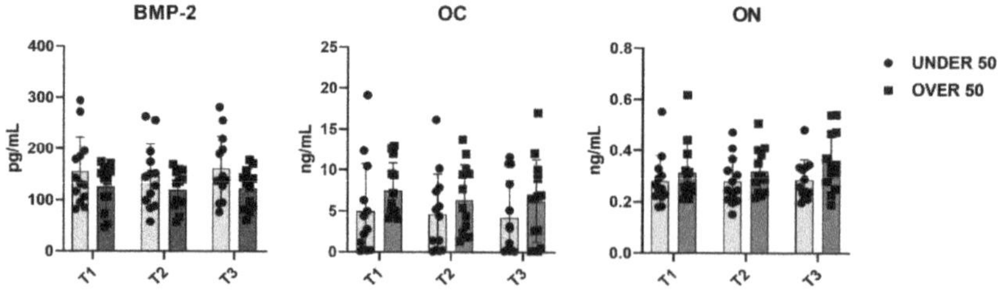


**Figure 4.** The average release of bone morphogenetic protein-2 (BMP-2), osteocalcin (OC) and osteonectin (ON) reported for all analyzed times (T1: 24 h; T2: 10 days; and T3: 28 days) in the UNDER 50 and OVER 50 groups. The results are expressed as the mean ± standard deviation of data obtained by duplicate analyses, performed using Prism 5.0 software (GraphPad). The statistical analyses were performed using a nonparametric one-way ANOVA (Kruskal–Wallis test) and post hoc Mann–Whitney *U* (MWU) test (adjusted by Bonferroni correction for multiple tests).

To evaluate the overall release of the three factors after 28 days, a comparative analysis was carried out. The average differences among patients and between the UNDER 50 and OVER 50 groups were not statistically significant. The averages of the released amounts at the three time points and for all samples, divided according to the two groups, UNDER 50 and OVER 50, are reported in Figure 5. As described in the above analysis, the BMP-2 and ON showed comparable releases in the two groups, whereas only the OC showed a difference. The release was higher in the OVER 50 group, even if the absence of statistical significance was confirmed.

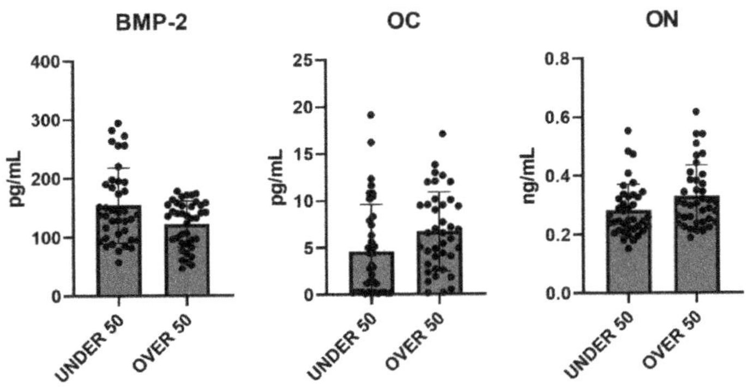


**Figure 5.** The average cumulative release of bone morphogenetic protein-2 (BMP-2), osteocalcin (OC) and osteonectin (ON) considering all samples from the UNDER 50 and OVER 50 groups at all analyzed times. The results are expressed as the mean ± standard deviation of data obtained by duplicate analyses, performed using Prism 5.0 software (GraphPad). The statistical analysis was performed using a nonparametric one-way ANOVA (Kruskal–Wallis test) and post hoc Mann–Whitney *U* (MWU) test (adjusted by Bonferroni correction for multiple tests).

The last decades have been characterized by a great interest in biomaterials for bone regeneration. After tooth extraction, bone remodeling invariably leads to a loss of bone volume. A variety of biomaterials, spanning from synthetic products to autologous or heterologous substitutes, are currently used after the tooth extraction to minimize this phenomenon [43].

It is accepted that the ideal graft material should be osteogenic, osteoinductive and osteoconductive at the same time [9]. From this perspective, only autologous bone meets all of these requirements [14,44]. Other widely used graft materials, such as heterologous bone or alloplastic materials grafts, can have excellent osteoconductive properties even if without osteoinduction [45].

In this scenario, the possibility of using extracted teeth as an autologous graft material in the form of demineralized dentin matrix (DDM) has come to the fore. The first reason for using DDM in bone regeneration is represented by their similarities in embryonic origin and chemical composition [46].

A review article published by Zhang et al. [47] compared DDM to other graft materials highlighting some peculiarities. In addition to the biocompatibility, the authors highlight the higher acceptance of DDM by patients compared to allografts from cadavers. Another advantage that also contributes to patients' acceptance is represented by the availability of dentin, eliminating the need for bone harvesting from other sites. The authors also conclude that autologous dentin is as effective as autologous bone in bone regeneration.

Like autologous bone, DDM is cost effective, potentially reducing the overall treatment cost compared to other bone substitutes.

In this study, we confirmed the release of BMP-2, OC and ON from ground dentin. Similar to other bone substitutes, demineralized dentin matrix (DDM) primarily functions as a scaffold [37]. However, in addition to its osteoconductive properties, the presence and release of BMP-2 endow it with osteoinductive capabilities, promoting the differentiation of mesenchymal stem cells into osteoblasts and subsequent bone matrix deposition. Furthermore, no significant differences in growth factor release were observed between the UNDER 50 and OVER 50 groups. Initially, we hypothesized that the potential of dentin to provide growth factors might decrease with age. This hypothesis was based on the relative increase in the amounts of secondary and tertiary dentin compared to primary dentin over a lifetime. However, our results suggest that age does not affect the regenerative potential of dentin. This can be explained by the fact that secondary dentin forms as a secondary layer along the entire pulp–dentin border throughout life, without replacing the primary dentin, and tertiary dentin is produced only in localized areas in response to harmful stimuli. Consequently, teeth, and specifically dentin, can be regarded as a reservoir of molecules beneficial for regeneration, maintaining their stability throughout an individual's lifetime. This finding further supports the use of autologous dentin, regardless of a patient's age, whenever possible.

Different devices are available on the market for grinding extracted teeth. A study conducted in 2023 by Dłucik R et al. [48] compared the efficacy of three DDMs obtained from different medical devices on the market (Tooth Transformer Smart, Smart Dentin Grinder and BonMaker). Despite some variations in structures and chemical compositions of the granules, the authors concluded that all of the devices could generate particles with adequate characteristics for bone regeneration. The present study was conducted using a patented medical device specifically designed to convert extracted teeth into DDM. Differently from other devices, the overall preparation procedure takes only a few minutes [19], making it particularly suitable for clinical practice. Moreover, since no weak acids have been used in the postground treatment, only a partial demineralization could be expected. Given the high mineral content of teeth, demineralizing protocols have been recently recommended to preserve autologous growth factors, including BMP-2, and to facilitate their sustained release over time [19,49]. Demineralizing protocols typically involve the use of weak acids, such as hydrochloric acid (HCl) or ethylenediaminetetraacetic acid (EDTA). This process removes the mineral content, exposing the organic matrix, which primarily consists of collagen and other noncollagenic molecules, including BMPs. Demineralized tooth granules may exhibit enhanced osteoinductive properties due to the exposure of growth factors such as BMPs compared to nondemineralized granules, although no specific literature data currently support this claim. Nonetheless, it has been observed that the demineralization process can render dentin remarkably similar to natural bone in terms of surface chemical composition [50].

While the biochemical study confirms the osteoinductive potential of dentin granules, the initial hypothesis regarding age-related influence was not supported, since the statistical analysis revealed no correlation between interindividual variability and patient's age. These findings call for alternative explanations. Firstly, factors related to bone metabolism might influence the mineral and organic content of the teeth at the time of their development; secondly, unknown aspects of patients' medical history or metabolism could play a role even after tooth development. Further studies involving larger populations should more thoroughly investigate various patient-related factors, such as the osteometabolic profile, which may influence the content and release of growth factors from dentin.

This study has some limitations, including the relatively small size of the study population and the lack of prior laboratory data on patients' bone metabolism, which would help investigate factors that might influence the content of growth factors in dentin. Additionally, the inclusion of different teeth with varying eruption ages may have introduced variability into the results.

While recent research has concentrated on characterizing dentin particles based on various grinding and postgrinding processes, future studies need to explore the biological activity of dentin grafts. This aspect is crucial as it distinguishes dentin grafts from other graft materials, such as xenografts and alloplastic bone substitutes. Once the potential of dentin grafts is fully elucidated, extracted teeth will justifiably be regarded as a valuable biological resource for bone regeneration procedures. For the same reason, it will be fundamental to investigate the possibility of storing ground teeth for future use with the same patient.

## 4. Conclusions

The results of this study indicate that BMP-2, OC and ON are effectively released by dentine granules over time. The statistical analysis demonstrated that the release of growth factors is not affected by age, though some variability exists among the patients, particularly for the OC release. These findings confirm DDM as an alternative to traditional biomaterials for bone regeneration procedures. Considering the limitations of this study, further research with larger sample sizes is warranted to confirm the findings and explore potential relationships with other patient-related factors.

**Author Contributions:** Conceptualization, G.M., M.L. and A.D.B.; methodology, G.M., A.M. and G.S.; investigation, G.M., G.S., L.L. and A.D.B.; data curation, G.M., A.M. and A.D.B.; writing—original draft preparation, G.M., M.L. and A.M.; writing—review and editing, A.S.d. and M.L.; supervision, L.L.; project administration, A.S.d. and A.D.B. All authors have read and agreed to the published version of the manuscript.

**Funding:** This research received no external funding.

**Institutional Review Board Statement:** The study was conducted according to the guidelines of the Declaration of Helsinki and approved by the Ethics Committee of Policlinico Umberto I of Rome, Italy (protocol code: 5456/2019, 25 July 2019).

**Informed Consent Statement:** Informed consent was obtained from all subjects involved in the study.

**Data Availability Statement:** The data presented in this study are available upon request from the corresponding author.

**Acknowledgments:** The Authors thank KometaBio (Fort Lee, NJ, USA), who kindly provided the device Smart Dentin Grinder™ and Alessia Baseggio Conrado for kind review of statistics.

**Conflicts of Interest:** The authors declare no conflicts of interest.

## References


1. Dahlin, C.; Linde, A.; Gottlow, J.; Nyman, S. Healing of bone defects by guided tissue regeneration. *Plast. Reconstr. Surg.* **1988**, *81*, 672–676. [CrossRef] [PubMed]
2. Amini, A.R.; Laurencin, C.T.; Nukavarapu, S.P. Bone tissue engineering: Recent advances and challenges. *Crit. Rev. Biomed. Eng.* **2012**, *40*, 363–408. [CrossRef] [PubMed] [PubMed Central]
3. Vittorini Orgeas, G.; Clementini, M.; De Risi, V.; de Sanctis, M. Surgical techniques for alveolar socket preservation: A systematic review. *Int. J. Oral. Maxillofac. Implant.* **2013**, *28*, 1049–1061. [CrossRef] [PubMed]
4. Haugen, H.J.; Lyngstadaas, S.P.; Rossi, F.; Perale, G. Bone grafts: Which is the ideal biomaterial? *J. Clin. Periodontol.* **2019**, *46* (Suppl. S21), 92–102. [CrossRef] [PubMed]
5. Al-Moraissi, E.A.; Alkhutari, A.S.; Abotaleb, B.; Altairi, N.H.; Del Fabbro, M. Do osteoconductive bone substitutes result in similar bone regeneration for maxillary sinus augmentation when compared to osteogenic and osteoinductive bone grafts? A systematic review and frequentist network meta-analysis. *Int. J. Oral. Maxillofac. Surg.* **2020**, *49*, 107–120. [CrossRef] [PubMed]
6. Sanz, M.; Dahlin, C.; Apatzidou, D.; Artzi, Z.; Bozic, D.; Calciolari, E.; De Bruyn, H.; Dommisch, H.; Donos, N.; Eickholz, P.; et al. Biomaterials and regenerative technologies used in bone regeneration in the craniomaxillofacial region: Consensus report of group 2 of the 15th European Workshop on Periodontology on Bone Regeneration. *J. Clin. Periodontol.* **2019**, *46* (Suppl. 21), 82–91. [CrossRef] [PubMed]
7. Wang, B.; Feng, C.; Liu, Y.; Mi, F.; Dong, J. Recent advances in biofunctional guided bone regeneration materials for repairing defective alveolar and maxillofacial bone: A review. *Jpn. Dent. Sci. Rev.* **2022**, *58*, 233–248. [CrossRef] [PubMed] [PubMed Central]
8. Kyyak, S.; Pabst, A.; Heimes, D.; Kämmerer, P.W. The Influence of Hyaluronic Acid Biofunctionalization of a Bovine Bone Substitute on Osteoblast Activity In Vitro. *Materials* **2021**, *14*, 2885. [CrossRef] [PubMed] [PubMed Central]
9. Giannoudis, P.V.; Chris Arts, J.J.; Schmidmaier, G.; Larsson, S. What should be the characteristics of the ideal bone graft substitute? *Injury* **2011**, *42* (Suppl. 2), S1–S2. [CrossRef] [PubMed]
10. Misch, C.M. Autogenous Bone is Still the Gold Standard of Graft Materials in 2022. *J. Oral Implant.* **2022**, *48*, 169–170. [CrossRef] [PubMed]
11. Bernard, G.W. Healing and repair of osseous defects. *Dent. Clin. N. Am.* **1991**, *35*, 469–477. [CrossRef] [PubMed]
12. Cushing, M. Autogenous red marrow grafts: Their potential for induction of osteogenesis. *J. Periodontol.* **1969**, *40*, 492–497. [CrossRef] [PubMed]
13. Gamradt, S.C.; Lieberman, J.R. Bone graft for revision hip arthroplasty: Biology and future applications. *Clin. Orthop. Relat. Res.* **2003**, *417*, 183–194. [CrossRef] [PubMed]
14. LeGeros, R.Z. Properties of osteoconductive biomaterials: Calcium phosphates. *Clin. Orthop. Relat. Res.* **2002**, *395*, 81–98. [CrossRef] [PubMed]
15. Mastrangelo, F. New Implant Materials. *Materials* **2023**, *16*, 4525. [CrossRef] [PubMed]
16. Ferraz, M.P. Bone Grafts in Dental Medicine: An Overview of Autografts, Allografts and Synthetic Materials. *Materials* **2023**, *16*, 4117. [CrossRef] [PubMed] [PubMed Central]

17. Von Arx, T.; Hardt, N.; Wallkamm, B. The TIME technique: A new method for localized alveolar ridge augmentation prior to placement of dental implants. *Int. J. Oral. Maxillofac. Implant.* **1996**, *11*, 387–394. [PubMed]
18. Tang, G.; Liu, Z.; Liu, Y.; Yu, J.; Wang, X.; Tan, Z.; Ye, X. Recent Trends in the Development of Bone Regenerative Biomaterials. *Front. Cell Dev. Biol.* **2021**, *9*, 665813. [CrossRef] [PubMed] [PubMed Central]
19. Cervera-Maillo, J.M.; Morales-Schwarz, D.; Morales-Melendez, H.; Mahesh, L.; Calvo-Guirado, J.L. Autologous Tooth Dentin Graft: A Retrospective Study in Humans. *Medicina* **2021**, *58*, 56. [CrossRef] [PubMed] [PubMed Central]
20. Mazzucchi, G.; Lollobrigida, M.; Lamazza, L.; Serafini, G.; Di Nardo, D.; Testarelli, L.; De Biase, A. Autologous Dentin Graft after Impacted Mandibular Third Molar Extraction to Prevent Periodontal Pocket Formation—A Split-Mouth Pilot Study. *Materials* **2022**, *15*, 1431. [CrossRef] [PubMed] [PubMed Central]
21. Kim, Y.K.; Kim, S.G.; Byeon, J.H.; Lee, H.J.; Um, I.U.; Lim, S.C.; Kim, S.Y. Development of a novel bone grafting material using autogenous teeth. *Oral Surg. Oral Med. Oral Pathol. Oral Radiol. Endodiol.* **2010**, *109*, 496–503. [CrossRef] [PubMed]
22. Bang, G.; Urist, M.R. Bone induction in excavation chambers in matrix of decalcified dentin. *Arch. Surg.* **1967**, *94*, 781–789. [CrossRef] [PubMed]
23. Kawai, T.; Urist, M.R. Bovine tooth-derived bone morphogenetic protein. *J. Dent. Res.* **1989**, *68*, 1069–1074. [CrossRef] [PubMed]
24. Kim, Y.K.; Kim, S.G.; Oh, J.S.; Jin, S.C.; Son, J.S.; Kim, S.Y.; Lim, S.Y. Analysis of the inorganic component of autogenous tooth bone graft material. *J. Nanosci. Nanotechnol.* **2011**, *11*, 7442–7445. [CrossRef] [PubMed]
25. Morotome, Y.; Goseki-Sone, M.; Ishikawa, I.; Oida, S. Gene expression of growth and differentiation factors-5, -6, and -7 in developing bovine tooth at the root forming stage. *Biochem. Biophys. Res. Commun.* **1998**, *244*, 85–90. [CrossRef] [PubMed]
26. Pang, K.M.; Um, I.W.; Kim, Y.K.; Woo, J.M.; Kim, S.M.; Lee, J.H. Autogenous demineralized dentin matrix from extracted tooth for the augmentation of alveolar bone defect: A prospective randomized clinical trial in comparison with anorganic bovine bone. *Clin. Oral. Implant. Res.* **2017**, *28*, 809–815. [CrossRef] [PubMed]
27. Kim, Y.K.; Kim, S.G.; Yun, P.Y.; Yeo, I.S.; Jin, S.C.; Oh, J.S.; Kim, H.J.; Yu, S.K.; Lee, S.Y.; Kim, J.S.; et al. Autogenous teeth used for bone grafting: A comparison with traditional grafting materials. *Oral Surg. Oral Med. Oral Pathol. Oral Radiol.* **2014**, *117*, e39–e45. [CrossRef] [PubMed]
28. Kozuma, W.; Kon, K.; Kawakami, S.; Bobothike, A.; Iijima, H.; Shiota, M.; Kasugai, S. Osteoconductive potential of a hydroxyapatite fiber material with magnesium: In vitro and in vivo studies. *Dent. Mater. J.* **2019**, *38*, 771–778. [CrossRef] [PubMed]
29. Kim, Y.K.; Lee, J.; Um, I.W.; Kim, K.W.; Murata, M.; Akazawa, T.; Mitsugi, M. Tooth-derived bone graft material. *J. Korean Assoc. Oral Maxillofac. Surg.* **2013**, *39*, 103–111. [CrossRef] [PubMed] [PubMed Central]
30. Butler, W.T.; Ritchie, H. The nature and functional significance of dentin extracellular matrix proteins. *Int. J. Dev. Biol.* **1995**, *39*, 169–179. [PubMed]
31. Ike, M.; Urist, M.R. Recycled dentin root matrix for a carrier of recombinant human bone morphogenetic protein. *J. Oral Implant.* **1998**, *24*, 124–132. [CrossRef] [PubMed]
32. Urist, M.R.; Strates, B.S. The classic: Bone morphogenetic protein. *Clin. Orthop. Relat. Res.* **2009**, *467*, 3051–3062. [CrossRef] [PubMed] [PubMed Central]
33. Bessho, K.; Tanaka, N.; Matsumoto, J.; Tagawa, T.; Murata, M. Human dentin-matrix-derived bone morphogenetic protein. *J. Dent. Res.* **1991**, *70*, 171–175. [CrossRef] [PubMed]
34. Bronckers, A.L.; Price, P.A.; Schrijvers, A.; Bervoets, T.J.; Karsenty, G. Studies of osteocalcin function in dentin formation in rodent teeth. *Eur. J. Oral Sci.* **1998**, *106*, 795–807. [CrossRef] [PubMed]
35. Papagerakis, P.; Berdal, A.; Mesbah, M.; Peuchmaur, M.; Malaval, L.; Nydegger, J.; Simmer, J.; Macdougall, M. Investigation of osteocalcin, osteonectin, and dentin sialophosphoprotein in developing human teeth. *Bone* **2002**, *30*, 377–385. [CrossRef] [PubMed]
36. Rijal, G.; Shin, H.I. Human tooth-derived biomaterial as a graft substitute for hard tissue regeneration. *Regen. Med.* **2017**, *12*, 263–273. [CrossRef] [PubMed]
37. Kim, K.W. Bone Induction by Demineralized Dentin Matrix in Nude Mouse Muscles. *Maxillofac. Plast. Reconstr. Surg.* **2014**, *36*, 50–56. [CrossRef] [PubMed] [PubMed Central]
38. Nampo, T.; Watahiki, J.; Enomoto, A.; Taguchi, T.; Ono, M.; Nakano, H.; Yamamoto, G.; Irie, T.; Tachikawa, T.; Maki, K. A new method for alveolar bone repair using extracted teeth for the graft material. *J. Periodontol.* **2010**, *81*, 1264–1272. [CrossRef] [PubMed]
39. Valdec, S.; Pasic, P.; Soltermann, A.; Thoma, D.; Stadlinger, B.; Rücker, M. Alveolar ridge preservation with autologous particulated dentin-a case series. *Int. J. Implant. Dent.* **2017**, *3*, 2. [CrossRef] [PubMed] [PubMed Central]
40. Kim, Y.; Lee, J.K.; Kim, K.; Um, I.; Murata, M. Healing mechanism and clinical application of autogenous tooth bone graft material. In *Advances in Biomaterials Science and Biomedical Applications*; InTech: London, UK, 2013; pp. 405–436.
41. Calvo-Guirado, J.L.; Ballester-Montilla, A.; NDe Aza, P.; Fernández-Domínguez, M.; Alexandre Gehrke, S.; Cegarra-Del Pino, P.; Mahesh, L.; Pelegrine, A.A.; Aragoneses, J.M.; Maté-Sánchez de Val, J. Particulated, Extracted Human Teeth Characterization by SEM-EDX Evaluation as a Biomaterial for Socket Preservation: An in vitro Study. *Materials* **2019**, *12*, 380. [CrossRef] [PubMed] [PubMed Central]
42. Memè, L.; Strappa, E.M.; Monterubbianesi, R.; Bambini, F.; Mummolo, S. SEM and FT-MIR Analysis of Human Demineralized Dentin Matrix: An In Vitro Study. *Appl. Sci.* **2022**, *12*, 1480. [CrossRef]

43. Nasr, H.F.; Aichelmann-Reidy, M.E.; Yukna, R.A. Bone and bone substitutes. *Periodontol. 2000* **1999**, *19*, 74–86. [CrossRef] [PubMed]
44. Silva, F.M.; Cortez, A.L.; Moreira, R.W.; Mazzonetto, R. Complications of intraoral donor site for bone grafting prior to implant placement. *Implant. Dent.* **2006**, *15*, 420–426. [CrossRef] [PubMed]
45. Chavda, S.; Levin, L. Human Studies of Vertical and Horizontal Alveolar Ridge Augmentation Comparing Different Types of Bone Graft Materials: A Systematic Review. *J. Oral. Implant.* **2018**, *44*, 74–84. [CrossRef] [PubMed]
46. Murata, M.; Akazawa, T.; Mitsugi, M.; Um, I.W.; Kim, K.W.; Kim, Y.K. Human Dentin as Novel Biomaterial for Bone Regeneration. *Biomater. Phys. Chem.* **2011**, *14*, 127–140.
47. Zhang, S.; Li, X.; Qi, Y.; Ma, X.; Qiao, S.; Cai, H.; Zhao, B.C.; Jiang, H.B.; Lee, E.S. Comparison of Autogenous Tooth Materials and Other Bone Grafts. *Tissue Eng. Regen. Med.* **2021**, *18*, 327–341. [CrossRef] [PubMed] [PubMed Central]
48. Dłucik, R.; Orzechowska-Wylęgała, B.; Dłucik, D.; Puzzolo, D.; Santoro, G.; Micali, A.; Testagrossa, B.; Acri, G. Comparison of clinical efficacy of three different dentin matrix biomaterials obtained from different devices. *Expert Rev. Med. Devices* **2023**, *20*, 313–327, Erratum in: *Expert. Rev. Med. Devices* **2023**, *20*, 883. [CrossRef] [PubMed]
49. Boden, S.D.; Liu, Y.; Hair, G.A.; Helms, J.A.; Hu, D.; Racine, M.; Nanes, M.S.; Titus, L. LMP-1, a LIM-domain protein, mediates BMP-6 effects on bone formation. *Endocrinology* **1998**, *139*, 5125–5134. [CrossRef] [PubMed]
50. Minetti, E.; Palermo, A.; Malcangi, G.; Inchingolo, A.D.; Mancini, A.; Dipalma, G.; Inchingolo, F.; Patano, A.; Inchingolo, A.M. Dentin, Dentin Graft, and Bone Graft: Microscopic and Spectroscopic Analysis. *J. Funct. Biomater.* **2023**, *14*, 272. [CrossRef] [PubMed] [PubMed Central]

Journal of
*Functional*
*Biomaterials*

MDPI

*Review*

# The Integration of Gold Nanoparticles into Dental Biomaterials as a Novel Approach for Clinical Advancement: A Narrative Review

**Saharat Jongrungsomran [1], Dakrong Pissuwan [2], Apichai Yavirach [1], Chaiy Rungsiyakull [3] and Pimduen Rungsiyakull [1,*]**

[1] Department of Prosthodontics, Faculty of Dentistry, Chiang Mai University, Chiang Mai 50200, Thailand; saharat_j@cmu.ac.th (S.J.); apichai.y@cmu.ac.th (A.Y.)
[2] Nanobiotechnology and Nanobiomaterials Research Laboratory, School of Materials Science and Innovation, Faculty of Science, Mahidol University, Bangkok 10400, Thailand; dakrong.pis@mahidol.ac.th
[3] Department of Mechanical Engineering, Faculty of Engineering, Chiang Mai University, Chiang Mai 50200, Thailand; chaiy.rungsiyakull@cmu.ac.th
* Correspondence: pimduen.rungsiyakull@cmu.ac.th; Tel.: +66-81-950-2400

**Abstract:** Gold nanoparticles (AuNPs) have gained significant attention in the biomedical field owing to their versatile properties. AuNPs can be customized by modifying their size, shape and surface characteristics. In recent years, extensive research has explored the integration of AuNPs into various dental materials, including titanium, polymethylmethacrylate (PMMA) and resin composites. This review aims to summarize the advancements in the application of modified AuNPs in dental materials and to assess their effects on related cellular processes in the dental field. Relevant articles published in English on AuNPs in association with dental materials were identified through a systematic search of the PubMed/MEDLINE, Embase, Scopus and ScienceDirect databases from January 2014 to April 2024. Future prospects for the utilization of AuNPs in the field of dentistry are surveyed.

**Keywords:** gold nanoparticles; dental biomaterial; antimicrobial activity; osteogenic activity; cell adhesion; cell proliferation; cell differentiation

## 1. Introduction

Nanotechnology has greatly enhanced modern healthcare, resulting in improved patient well-being and quality of life. It has facilitated the advancement of novel and more efficient pharmaceuticals, medical procedures and diagnostic instruments. Nanotechnology has numerous applications in dentistry and is crucial to the development of advanced dental equipment used in treatment. Nanoparticles (NPs) can be precisely manipulated because of their size, which is at the nanoscale level (less than 100 nm), as well as their morphology. The interaction of NPs at the molecular level is more efficient than that of micro- or macro-sized particles. NPs can be easily controlled and used in a wide range of applications because of their large surface area. The prerequisites for utilizing NPs include being biologically inactive, not causing cancer, possessing adequate mechanical durability and being resilient to the internal conditions of the body. There are several widely recognized types of nanoparticles, including carbon-based nanoparticles, metal nanoparticles, ceramic nanoparticles, polymeric nanoparticles and lipid-based nanoparticles; these can be distinguished based on their features. In recent years, numerous researchers have highlighted the significant influence of NPs in the fields of tissue engineering and regenerative medicine [1,2].

AuNPs are commonly used in the medical field, in a variety of applications. They have facilitated the development of effective drugs, medical techniques and diagnostic tools. AuNPs can be controlled at the nanoscale level, including their shape and structure, and have recently been reported to show favorable antimicrobial performance, anticancer effects and antioxidant and anti-inflammatory activities with less toxicity than other metal

NPs. Their relative biocompatibility and easy synthesis make them suitable for various biomedical applications, such as drug delivery, imaging and therapy. The factors that affect the properties are the size, structure and surface modifications [1,2]. Moreover, many studies have shown that AuNPs can promote stem cell differentiation, especially in mesenchymal stem cells (MSCs) [3–8]. The surface of AuNPs can be stabilized by using various molecules [9–11]. Various studies have shown that AuNPs are useful in biomedical applications, including target cell therapy, tissue engineering and regenerative medicine [1,2,8,12–14]. Consequently, AuNPs have been engineered in various forms to enhance their efficacy in these applications.

Firstly, the size of AuNPs is dominantly dependent on the cellular uptake and biological distribution [15–17]. The uptake and penetration of smaller AuNPs in targeted cells are influenced by their size, which makes them more prone to aggregation in the target cells [18]. Smaller AuNPs usually exhibit higher catalytic efficiency due to their increased surface area-to-volume ratio, providing more active sites for catalytic reactions [19]. The antimicrobial properties of nanoparticles, which can be attributed to their extremely small size, which is up to 250 times smaller than that of bacteria, result in a higher cellular uptake and excess reactive oxygen species (ROS). However, it is important to note that 4 nm spherical AuNPs have a greater tendency to cause cytotoxicity and unfavorable adipogenic differentiation instead of osteogenic differentiation compared to 40 nm AuNPs [8]. AuNPs larger than 80 nm may have difficulty penetrating the cell membrane, as with smaller AuNPs, but they can adhere to bacterial cells to increase tension in the membrane and lead to cell rupture [20]. Secondly, AuNPs adopt a variety of shapes, each with unique properties and applications. In addition to their size, the rate at which AuNPs are taken up by cells is also influenced by their shape, which ultimately affects the number of AuNPs that accumulate inside cells [8,16]. Numerous shapes of AuNPs can be produced, but the most prevalent are spherical and rod-shaped, which are relatively straightforward to synthesize. According to recent findings, spherical AuNPs have demonstrated a greater ability to enhance osteogenic activity and calcium deposition in human MSCs (hMSCs) compared to rod-shaped AuNPs [8]. Another type, star-shaped AuNPs, have been employed in surface-enhanced Raman scattering (SERS), photothermal therapy (PTT) and biological imaging applications. AuNPs can act as either the core or shell component in core–shell structures, resulting in increased stability, high biocompatibility and the ability to be functionalized for drug delivery systems [21]. Lastly, surface modifications play an important role in the application of the target cell [16]. The surface alteration not only improves cell activities but also stabilizes nanoparticles, allowing them to diffuse in bodily fluids [8]. AuNPs coated with specific molecules can attach to specific receptors, resulting in active targeting. Furthermore, passive targeting is dependent on the efficiency of cellular absorption [16].

Biomaterials have been used in various applications in dentistry. These materials exhibit promising mechanical properties and biocompatibility in the oral cavity. However, there remains the potential to enhance their performance and longevity. Previous investigations revealed alterations made to dental materials such as titanium [22], PMMA [23] and resin composites by the incorporation of AuNPs [24]. The effective combination of AuNPs with zirconia or polyetheretherketone (PEEK) has also been demonstrated [25]. Titanium has numerous applications in dentistry, exhibiting excellent success rates and eliciting positive biological reactions upon contact with living tissues. Titanium is commonly used in dental implants and its components [26]. Polymeric materials such as PMMA and resin composites have been developed for use in temporary fixed or removable dental prostheses [20,27]. Resin composites have also emerged as materials for direct restoration and modern luting cements [28,29]. Furthermore, nanotechnology modifications such as coatings, sputtering or even mixing on dental materials have been investigated for improving mechanical and biological properties and allowing for biomedical applications [1,2]. These biomaterials are essential to many dental operations and offer a range of options for clinicians to choose from based on the specific requirements and desired outcomes.

As previously noted, AuNPs have proven to be highly effective in a range of biomedical applications, including drug delivery systems [30], enhancing the effectiveness of antimicrobial and anticancer drugs [31,32], photothermal cancer therapy [33] and diagnostic assays [34,35]. Although numerous studies have investigated the use of AuNPs in other fields of biomedicine, there is a lack of research on the long-term biocompatibility and mechanical stability of AuNPs when incorporated into dental materials. In addition, the impact of surface changes on the clinical results of these materials has not been extensively investigated. Given their success in these areas, it is an attractive prospect to investigate their potential use in dental applications. Therefore, this article aims to summarize the properties of AuNPs with various modifications, when used for dental materials, and review their effects on related cells in the field of dentistry.

## 2. Methodology

An electronic search was performed by using the PubMed/Medline, Embase, Scopus and ScienceDirect databases with the following combinations of keywords: ("Gold nanoparticles" OR ("Gold" AND "Nanoparticle") AND ("Dental" OR "Dentistry" OR "Dental applications"). The search was limited to English-language articles published from January 2014 to April 2024. The electronic results included 512 articles; we eliminated duplicate articles via software for managing references (EndNote 20; Thomson Reuters, Philadelphia, PA, USA) for a total of 358 articles. Following the initial search, the titles and abstracts were reviewed to find relevant papers. Studies that were not conducted in the dentistry field and did not involve the use of AuNPs were excluded. Additionally, articles obtained by electronic hand-searching were included. Finally, we chose 113 articles to include in this comprehensive review (Figure 1). The search terms are shown in Table 1.

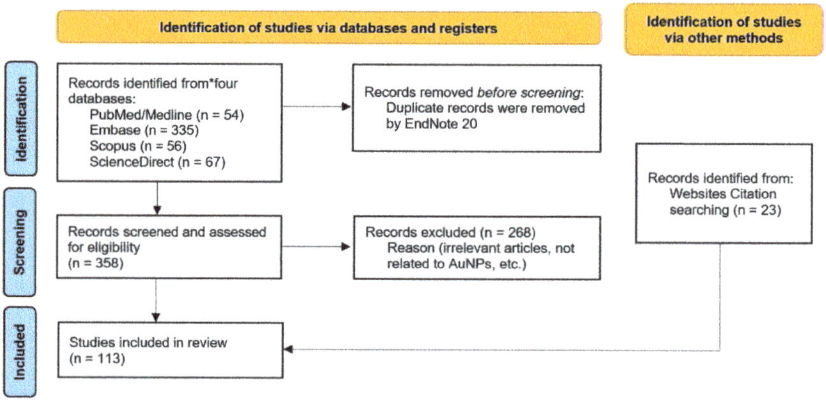


**Figure 1.** Flow diagram of search procedure.

**Table 1.** Search strategy for four databases (PubMed, Embase, Scopus, ScienceDirect).

| Database | Search Strategy |
|---|---|
| PubMed/Medline | ("dental health services" [MeSH Terms] OR "Dental" [Text Word]) AND ("gold" [MeSH Terms] OR "gold" [Text Word]) AND ("Nanoparticles" [MeSH Terms] OR "Nanoparticles" [Text Word]) |
| Embase | ('dental medicine' OR 'dental specialties' OR 'dental specialty' OR 'dental system' OR 'occupational dentistry' OR 'pathology, oral' OR 'specialties, dental' OR 'state dentistry' OR 'dentistry') AND ('Au nano-particle' OR 'Au nanoparticle' OR 'gold nano-particle' OR 'gold nanoparticles' OR 'nano gold' OR 'nano-Au' OR 'nanogold' OR 'nanoparticulate Au' OR 'nanoparticulate gold' OR 'gold nanoparticle') |
| Scopus | ("Dental application") AND ("Gold nanoparticle") |
| ScienceDirect | (dental applications) AND (gold AND nanoparticle) |

## 3. Results

### *3.1. Overview of AuNPs in the Biomedical Field*

AuNPs have been extensively utilized in various biomedical fields due to their unique properties and versatility. They are employed in cancer treatment, biomedical imaging, drug delivery systems, for the improvement of biomaterial properties and in other applications. Through biofunctionalization, AuNPs can be adapted for use in drug administration, vaccine development, sensing and imaging applications, demonstrating considerable potential in these domains. Furthermore, AuNPs play a critical role in tissue engineering and regenerative medicine, offering benefits such as tissue development and precise drug delivery. They can also be created with various forms and dimensions, which enhances their characteristics and potential uses [36]. AuNPs are plasmonic nanoparticles that appear to be less toxic. Because AuNPs are inert, stable and biocompatible, they have been frequently studied for their potential use in biosensing and drug delivery applications. The use of chemical molecules with antibacterial properties to coat AuNPs can offer significant advantages in terms of harnessing their benefits. Consequently, they are promising candidates for delivering targeted antibacterial actions [12]. The size, structure and surface modification of AuNPs have an impact on their characteristics. Furthermore, numerous studies have demonstrated that AuNPs, due to their ultra-small size, large surface-area-to-mass ratio and increased chemical reactivity, are useful in the biomedical field [1,2,8]. Research on AuNPs spans various biomedical fields including dental applications, which can be broadly categorized into three major areas: biological, optical and mechanical (Figure 2). A summary of the studies focusing on the use of AuNPs in biomedical fields is given in Table 2.

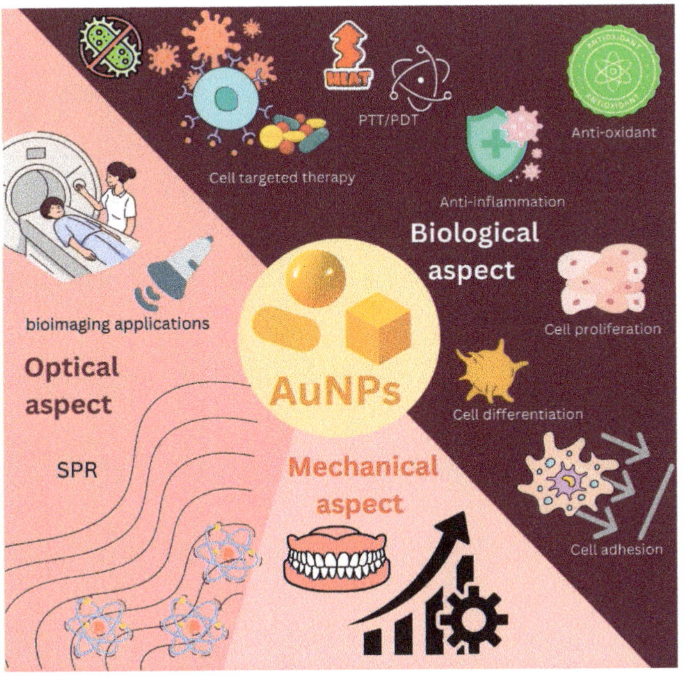


**Figure 2.** AuNPs in biomedical fields.

**Table 2.** Properties of AuNP-based nanomaterials in biomedical fields.

| Property | AuNP Molecule | | | Associated Cell or Biomarker Assay | Finding | Reference |
|---|---|---|---|---|---|---|
| | Size (nm) | Shape | Surface Modification | | | |
| Antimicrobial and antiplaque | 2 | Nanoclusters | Mesoporous silica nanoparticle-lysozyme-functionalized gold nanoclusters with kanamycin | *E. coli* | A mixed-matrix membrane coating controllably released an antimicrobial agent to instantly inhibit bacterial growth that was only triggered by bacteria. | [37] |
| | N/A | Spherical with core–shell structure | Core–shell magnetic nanoparticles | *S. mutans*, *E. coli* and *C. albicans* | MNP@Au decreased the adhesion of Gram-positive bacteria cells and fungal cells by 65% and Gram-negative bacteria cells by 45% without influencing the rheological and physicochemical properties of artificial saliva. | [38] |
| | AuNPs/CS = 16.21 nm AuNPs/ CS-Cur = 20.89 nm | Spherical | Chitosan (AuNPs/CS) Chitosan–curcumin (AuNPs/CS-Cur) | *S. mutans* and *E. coli* | AuNP/CS-Cur nanocomposites released curcumin in a pH-dependent manner, which may facilitate the drug to be delivered to the acidic bacterial infection environment. AuNP/CS-Cur possessed the characteristics of electrostatic targeting and photodynamic and photothermal antibacterial therapy; this would become an efficient and safe antibacterial nano-platform and provide new ideas for the treatment of bacterial infection. | [39] |
| | 5–15 | Spherical | None | *S. mutans* and *E. coli* | Green synthesized AuNPs exhibited significant antibacterial activity against *S. mutans* and *E. coli*, the same as the commercial ampicillin, in a dose-dependent trend. | [40] |
| | 45–70 | Spherical | None | *S. aureus* and *S. mutans* | Green synthesized AuNPs from wheat bran exhibited significant antibacterial activity against *S. aureus* and *S. mutans*. | [41] |
| | 18–35 | Spherical | None | *C. albicans*, *S. mutans*, *S. aureus* and *E. coli* | The antimicrobial property for AuNPs was assessed using a zone of inhibition test. They found that AuNPs exhibited excellent antimicrobial activity on *C. albicans* and intermediate antimicrobial potential on *S. mutans*, *S. aureus* and *E. coli*. | [42] |
| | 25, 60, 90 | N/A | None | *S. mutans*, *S. sanguinis* and *S. salivarius*. | The smallest size mediated the fast bactericidal activity and showed the most potent antibacterial activity. The 25 nm AuNP was more potent than chlorhexidine against evaluated standard species of *S.mutans*, *S.salivarius* and *S.sanguinis*. | [43] |
| | 43 | Spherical | None | *S. oralis* | Au NPs at 100 ppm and 50 ppm concentrations were equally as effective as the CHX against *S. oralis*. | [44] |
| | 20 | Spherical | Curcumin | *S. aureus*, *E. faecalis* and *C. albicans* | The antibacterial action of CuAUNPs was seen to be powerful against *S. aureus*, *E. faecalis* and *C. albicans* at a concentration of 100 μg/mL. | [45] |
| | 50 | Cube | None | *Candida species*; *C. albicans*, *C. glabrata* and *C. tropicalis* | The antifungal properties of gold nanocubes against *Candida species* isolates were greater than gold nanospheres and wires. | [46] |

**Table 2.** *Cont.*

| Property | AuNP Molecule | | | Associated Cell or Biomarker Assay | Finding | Reference |
|---|---|---|---|---|---|---|
| | Size (nm) | Shape | Surface Modification | | | |
| Antimicrobial and antiplaque | 32–41 | Nanocages | Antibiotic-loaded, antibody-conjugated, polydopamine (PDA); Staphylococcal protein A (Spa) and an antibiotic (daptomycin, ceftaroline, gentamicin and vancomycin) | *S. aureus*, methicillin-resistant *S. aureus* (MRSA) and *P. aeruginosa* | Daptomycin-loaded AuNCs that are linked to antibodies targeting two distinct lipoproteins of *S. aureus* successfully eradicate MRSA within a biofilm setting. The effectiveness of ceftaroline and vancomycin-loaded AuNCs, which were linked to anti-Spa antibodies, was observed to be lower compared to daptomycin-loaded AuNCs that were also linked to the same antibody. On the other hand, AuNCs loaded with gentamicin and linked to an antibody that targets a specific outer membrane protein were extremely successful in combating *P. aeruginosa* biofilms. The effectiveness of antibiotic-loaded, antibody-conjugated, PDA AuNCs relies on the capacity to induce both a fatal photothermal effect and the controlled release of a significant amount of antibiotic through heat. | [47] |
| | 30 | Colloidal | None | *S. mutans* | The combination application of low-temperature plasma and AuNPs resulted in substantial cellular rupture, leading to the release of intracellular constituents from several cells, while AuNPs alone did not showed any bactericidal effect. | [48] |
| | 8–12 | Nanoclusters | Methylene blue | *S. mutans* | Bovine serum albumin-capped gold nanoclusters conjugating with methylene blue demonstrated significant antibacterial effects against *S. mutans* when subjected to white-light LED irradiation for approximately 1 min. | [49] |
| Anti-inflammation and antioxidant | 45–70 | Spherical | None | DPPH assay | Green synthesized AuNPs from wheat bran possessed high antioxidant activity, which augmented in a dose-dependent manner. | [41] |
| | 20 | Spherical | Curcumin | DPPH assay and albumin denaturation assay | At a concentration of 50 μg/mL, CuAuNp demonstrated a maximum scavenging performance of 90.3% against DPPH. CuAuNP shown 79.6% more anti-inflammatory activity at a concentration of 50 μg/mL compared to the conventional medication diclofenac. | [45] |
| | 2–35 | Spherical | None | Protein (bovine serum albumin; BSA) denaturation inhibitory DPPH assay | Biosynthesized red sandal AuNP showed 83% (the highest) inhibitory activity of DPPH radicals at the highest concentration of 50 μg/mL. The highest inhibition and maximum protective activity of red sandal AuNP was 80.5% at a concentration of 50 μg/mL. AuNPs exhibited good antioxidant and anti-inflammatory properties. | [50] |
| | 30 | Spherical | Ursodeoxycholic acid (UDCA) | Nitric oxide (NO) test and enzyme-linked immunosorbent assay (ELISA) | The thermal effects of NIR-irradiated AuNP-UDCA inhibited the production of inflammatory cytokines from activated macrophages in vitro. The anti-inflammatory effects produced by GNP-UDCA under NIR irradiation were also shown in in vivo conditions using spinal-cord-injured rats. | [51] |
| | 70 | Spherical | None | DPPH assay | Biosynthesized AuNPs showed improved antioxidative action at a 1000 μg/mL concentration. | [52] |

**Table 2.** *Cont.*

| Property | AuNP Molecule | | | Associated Cell or Biomarker Assay | Finding | Reference |
|---|---|---|---|---|---|---|
| | Size (nm) | Shape | Surface Modification | | | |
| Anticancer activity | N/A | Spherical | *Siberian ginseng* aqueous extract | Rhodamine 123, H2DCFDA and dual AO/EtBr staining techniques | SG-AuNPs increased the reactive oxygen species and decreased the mitochondrial membrane potential, which induced apoptosis in melanoma cells, and it possessed an anticancer property. | [53] |
| | 50 | Spherical | None | Human breast cancer cell lines (MCF-7) | The most effective MCF-7 inhibition was seen when the concentration of AuNPs was 50 μg/mL and they were incubated for 4 h prior to irradiation with the lowest laser power of 0.002 W at a dosage of 0.61 J/cm$^2$ for 60 s. | [54] |
| Cell differentiation | 29–38 | Spherical | mPEG-SH, the cell-penetrating peptide RGD (RGDRGDRGDRGDPGC) and the mitochondria localization signal (MLALLGWWWFF-SRKKC) MLS@RGD-PEG | DPSCs ALP assay kit $Ca^{2+}$ assay kit ATP assay kit | Mitochondrial nanoprobes (MT-AuNPs/MLS@RGD-PEG-AuNPs) showed the potential molecular mechanisms of the accelerated differentiation of DPSCs, where the MMP was increased to induce more ATP production through the thermoplasmonic effect of AuNPs by in situ SERS, the results of which demonstrated that the expression of glucose and hydroxyproline increased within DPSCs during the cell differentiation process. | [55] |
| | 18 | Spherical | None | ALP, qRT-PCR of COLIα, RUNX2 and OCN | The addition of AuNPs boosted the behavior of hDPSCs on CPC. This improvement was observed in terms of increased cell adhesion (about double the amount of cell spreading) and proliferation, as well as enhanced osteogenic differentiation (approximately two to three times higher after 14 days). | [56] |
| Osteoinductive activity | 72–88 | Spherical | None | MTT (3-(4, 5-dimethylthiazol-2-yl)-2,5-diphenyltetrazolium bromide) assay | AuNPs produced showed exceptional stability in various blood components. Furthermore, the AuNPs were determined to be non-toxic when assessed for their compatibility with periodontal fibroblasts and erythrocytes, indicating their suitability for use in biomedical applications. When GNPs were exposed to MG-63 cell lines, there was a higher percentage of viable cells compared to the control group, indicating that AuNPs could induce bone formation. | [57] |
| | 28–32 | N/A | N-acetyl cysteine (NAC) | ALP activity | Gel-Ty hydrogels infused with G-NAC (Gel-Ty/G-NAC) exhibited adequate mechanical strength and biocompatibility to encapsulate and facilitate the proliferation of human adipose-derived stem cells (hASCs) across a three-day assessment. AuNPs may facilitate intracellular delivery of NAC. Furthermore, G-NAC facilitated osteodifferentiation both when incorporated into Gel-Ty and when applied directly to hASCs. The osteogenic effects were evidenced by the alkaline phosphatase (ALP) activity assay. | [58] |

**Table 2.** *Cont.*

| Property | AuNP Molecule | | | Associated Cell or Biomarker Assay | Finding | Reference |
|---|---|---|---|---|---|---|
| | Size (nm) | Shape | Surface Modification | | | |
| Imaging agent | 40 | N/A | poly (L-lysine) | Bright-field microscopy, confocal microscopy and quantification of the incorporation index<br>Viability assay–MTS assay<br>Micro-CT analysis | It is feasible to integrate the AuNP-PLL complex into DPSC and monitor the cellular behavior in a 3D analysis using micro-CT. Additionally, the inclusion of 0.2 mg/mL of AuNP-PLL does not disrupt the fundamental behavior of DPSC. | [59] |
| | N/A | N/A | None | Optical coherence tomography (OCT) | The addition of AuNPs into root canal irrigants enhances their optical opacity. Gold addition has been found to reduce micro-infiltration at the level of root canal walls, enhancing the adherence of filling materials to dentin. | [60] |

### 3.1.1. General Techniques for AuNPs' Synthesis and Surface Modification

In 1951, the Turkevich method, one of the most well-known methods for the synthesis of spherical AuNPs, was introduced. It relies on the reduction of $HAuCl_4$ by citrate in water [12]. AuNPs can be synthesized using a variety of methods, including physical, chemical and biological processes. One of the most promising approaches is the environmentally friendly "green synthesis", which has gained significant attention in recent years. The synthesis of AuNPs using plant extracts reduces chemical side effects and avoids creating toxic by-products. Various plant extracts can be used as a reducing agent, whereby $Au^{3+}$ is easily reduced to $Au^0$ [41,42,45,50,52,53,61].

Surface modifications of AuNPs play an important role in biomedical applications. Modified molecules such as ligands, proteins, drugs, antibodies and oligonucleotides are decorated on AuNPs' surface to specifically affect the target cells [16]. This modification cannot only improve the cell adhesion and proliferation capability but also stabilize the diffusion of nanoparticles in biological fluids [8]. Techniques for the surface modification of AuNPs can be divided into two major groups. The first group is physical interactions, including electrostatic and hydrophobic interactions. These interactions are simple and spontaneous processes; however, the interactions are unstable and responsive to various environmental conditions. Due to the electrostatic attraction of negatively charged AuNPs, drugs may bind covalently to AuNPs through amine groups and show better cell penetration [16,62]. On the contrary, positively charged AuNPs showed higher cellular uptake in hMSCs [7]. The second group is chemical interactions, which are more stable and less sensitive to environmental changes, although these interactions required complex processes. The cell targeting can involve both active and passive reactions. AuNPs decorated with specific molecules can bind to specific receptors on the target cell; this is called active targeting. Passive targeting depends on the effectiveness of cellular absorption, which is related to enhanced permeability and retention (EPR) effects [16,63]. In addition, selected molecules coated on AuNPs have a significant impact on targeting specific cells, which increases the mortality and DNA damage of targeted cells and results in fewer systemic side effects [19,64].

### 3.1.2. Impact of AuNPs' Size and Shape on the Cellular Uptake of Stem Cells

The size of AuNPs is a fundamental parameter that controls their properties and behaviors. It is a key consideration in the biofunctions and application of AuNPs because the size determines the cellular uptake and biological distribution [15–17]. The optimal size, between 25 and 50 nm, is more prone to aggregation in the target cells [18]. However, the rate of cellular uptake is not related to the size of AuNPs [16]. AuNPs adopt a variety of shapes, each with unique properties and applications. The cellular uptake rate is instead related to the shape of AuNPs, which significantly affects the number of AuNPs that accumulate in cells [8,16,46]. The most common shapes are spherical and rod-shaped; these are relatively simple to synthesize, have strong plasmonic properties and allow for precise tuning of the surface plasmon resonance (SPR) wavelength with the highest uptake rate. Although both spherical and rod-shaped AuNPs show a strong cellular uptake rate, spherical AuNPs can promote higher osteogenic induction, osteogenic differentiation, alkaline phosphatase (ALP) activity and calcium deposition of hMSCs than rod-shaped AuNPs [8]. Spherical AuNPs had a stimulatory effect on the proliferation rate and biocompatibility of dental pulp stem cells (DPSCs) derived from exfoliated deciduous teeth (SHED) [65]. Furthermore, cube-shaped AuNPs present a total of 12 edges, whereas nanospheres lack edges. It is plausible that the presence or absence of edges may influence cell adsorption activities [46].

#### 3.1.3. Impact of AuNPs' Size and Shape on Bacteria Present in the Mouth

The antimicrobial properties of nanoparticles can be attributed to their extremely small size, which can be up to 250 times smaller than that of bacteria. The small size of AuNPs can lead to electrostatic interactions between the Au atom and the negatively charged cell wall of the microbes [45]. Smaller AuNPs also have a higher surface-area-to-volume ratio, leading to stronger SPR effects and the absorption of shorter wavelengths of light. Moreover, smaller AuNPs typically exhibit higher catalytic efficiency, providing more active sites for catalytic reactions [19]. A 25 nm AuNP had more potent antibacterial action than chlorhexidine against evaluated standard species of *S. mutans*, *S. salivarius* and *S. sanguinis* [43]. However, some reports showed that a smaller size of spherical AuNPs (~4 nm) induced a higher cellular uptake and excess ROS than 40 nm AuNPs. In Ibrahim, et al. [66], 5 nm AuNPs showed greater cytotoxicity than 10 and 80 nm AuNPs. In addition, apoptosis and necrosis were activated through ROS generation. The surplus ROS induced higher cytotoxicity and unfavorable adipogenic differentiation rather than osteogenic differentiation [8]. Larger AuNPs, especially 50 nm, can readily penetrate cell membranes and access intracellular compartments, making them suitable for drug delivery and imaging applications [4]. Larger AuNPs, 80 to 100 nm, can adhere on bacterial cells, increasing tension in the membrane and leading to cell rupture [20]. The morphology of AuNPs predominantly comprises spherical and rod-shaped structures, which can be utilized in various applications. AuNPs with alternative geometries, such as star-shaped, have found applications in SERS, photothermal therapy and biological imaging. Furthermore, AuNPs can serve as either core or shell components in core–shell structures. A core–shell structure offers synergistic properties resulting from the combination of different materials, such as enhanced stability, excellent biocompatibility and the ability to be functionalized in drug delivery systems [21].

### *3.2. Application of AuNPs in Dental Biomaterials*

The incorporation of AuNPs into dental materials has great potential for improving dental treatments and patient outcomes (Figure 3). The dental biomaterial used within the human body should effectively prevent the development of biofilms by pathogens, minimizing the risk of severe consequences [67]. AuNPs exhibit distinctive characteristics, including antibacterial activity, enhancement of mechanical properties and targeted drug delivery capabilities, which make them helpful for enhancing the effectiveness of regenerative and tissue engineering. Moreover, researchers have investigated the application of AuNPs in regulating bacterial pathogenicity in the mouth without causing harm to cells, demonstrating their potential as substitutes for conventional antibiotics in the treatment of periodontal diseases. Coating with AuNPs had an inhibitory effect on the formation, maturation and viability of *S. aureus* biofilms [67]. Pure and modified surfaces of AuNPs with conventional antibiotics exhibit effective antibacterial activity against Gram-negative and Gram-positive bacteria through adhesion and penetration on the cell membrane. Moreover,

AuNPs show antiviral activity through interaction with viral surface proteins [19,68–72]. A review of each article about the use of modified AuNPs in dental materials is presented in Table 3.

**Table 3.** AuNPs' modification of dental materials.

| Material | Modification Technique on Materials | Modified Factors of AuNPs | | | Enhanced Activities | Findings | Reference |
|---|---|---|---|---|---|---|---|
| | | Size (nm) | Shape | Surface Modification | | | |
| Titanium | Magnetron sputtering | 8–26 | Rod | AuPt/PtAu | Osseointegration Antibacterial | AuPt/PtAu $TiO_2$ strongly affected the osteogenic performance of the hMSCs when using plasmonic photocatalysis and low-level laser therapy at 470 and 600 nm. | [6] |
| | Chemisorption: Gold-sulfur bonding (Au-S bond) | 30 | Spherical | (3-mercaptopropyl) trimethoxysilane ($SH(CH_2)_3Si(OCH_3)_3$, 3-MTPMS) | Osseointegration | AuNP-immobilized titanium implant showed higher values in osteogenic differentiation of human bone-marrow-derived MSCs and higher osseointegration parameters than an HA-titanium implant. | [14] |
| | Chemisorption: Gold-sulfur bonding (Au-S bond) on silanized Ti surface with (3-mercaptopropyl) trimethoxysilane ($SH(CH_2)_3Si(OCH_3)_3$, MTPMS) | 28 | Spherical | None | Osseointegration | Ti-AuNPs significantly enhanced the osteogenic differentiation in human adipose-derived stem cells (ADSCs) and influenced the osseous interface formation. | [22] |
| | Magnetron sputtering | 20–26 | Spherical | None | Antibacterial | Au@TNTs exhibit a preferable effect in restricting the growth of *E. coli* and *S. aureus* without cell cytotoxicity. | [73] |
| | Decomposition of $HAuCl_4$ by UV-light illumination | N/A | Spherical | None | Antibacterial and soft tissue healing promotion | An enhanced antibacterial effect against multispecies biofilm was realized via photocatalytic activity triggered by visible-light irradiation and generating a greater quantum yield of reactive oxygen species. AuNP decoration would also exert favorable effects on cell attachment, proliferation and migration fibroblast. | [74] |
| | Dipping technique (solution immersion) | N/A | N/A | Chitosan, DNA/c-myb | Osseointegration | The results from studying rat mandibles in vivo showed that titanium (Ti) implants coated with Ch-GNP/c-myb increased both the volume and density of newly created bone. Additionally, the inspection using microcomputed tomography revealed improved osseointegration between the dental implant and the bone, particularly in cases of ovariectomized osteoporosis. | [75] |
| | Decomposition of $HAuCl_4$ by UV-light illumination | 20 | Spherical | None | Antibacterial | AuNPs–TNTs showed excellent biocompatibility and exhibited significantly enhanced antibacterial activity against *P. gingivalis* via ROS generation under a simple ultrasound treatment. | [76] |
| | Immersion/gel coating with poly(lactic-co-glycolic acid) PGLA solution | 15–20 | Spherical | AntagomiR204 | Osseointegration in type 2 DM (in vivo study) | AuNP-antagomiR204 was released from the PGLA sheet and taken up by adherent BMSCs. This study showed the titanium implant promoting osseointegration in type 2 DM rats. | [77] |

**Table 3.** *Cont.*

| Material | Modification Technique on Materials | Modified Factors of AuNPs | | | Enhanced Activities | Findings | Reference |
|---|---|---|---|---|---|---|---|
| | | Size (nm) | Shape | Surface Modification | | | |
| Titanium | Magnetron sputtering | 30–38 | Rod | Tetracycline/ polycaprolactone (TC/PCL) | Antibacterial activity | AuNPs can produce remote-controlled tetracycline elution using near-infrared laser irradiation. | [78] |
| | Solution immersion | 10 | Spherical | None | Osseointegration | Photofunctionalized GNPs can highly improve the osteogenic capabilities with enhanced gene expression of osteogenic markers; Col-1, OPN and OCN; and ALP activity. | [79] |
| | Solution immersion | N/A | N/A | Chitosan, insulin growth factor binding protein-3 (IGFBP-3) | Osseointegration | Utilizing Ch-AuNPs conjugated with IGFBP-3 as a covering for titanium implants improved the process of bone formation and the integration of dental implants with the surrounding bone tissue against methylglyoxal-induced bone deterioration in a rat model. | [80] |
| PMMA | Au suspension in PMMA monomer | 57–82 | Spherical | None | Antibacterial and mechanical properties (microhardness) | PMMA/AuNPs showed a more significant antibiofilm effect against the monomicrobial biofilm formation of *C. albicans*, *S. mitis*, *S. aureus* and *E. coli* than PMMA. AuNPs slightly increased the Vicker's microhardness of PMMA. | [20] |
| | Physically combined in a ratio of 0.05 wt/wt % of AuNP powder | 10–20 | Spherical | None | Mechanical properties affecting dimensional stability | AuNP-modified heat-cured PMMA showed better dimensional stability and decrease in the artificial tooth movement than conventional PMMA. | [23] |
| | Au suspension in PMMA monomer | 11 | Spherical | None | Antifungal activity | PMMA/AuNP surfaces were smoother and more hydrophilic with negative charge than pure PMMA surfaces, which reduced the adhesion of *C. albicans*. AuNPs decreased the tensile strength of PMMA significantly. | [27] |
| | Au suspension in PMMA monomer with concentrations 0.05% and 0.2% | 45–65 | N/A | None | Mechanical properties affecting flexural strength | The flexural strength of PMMA with the addition of 0.05% AuNPs was significantly greater than with 0.20% AuNPs, whereas differences in sizes of AuNPs added to PMMA did not significantly affect its flexural strength. | [61] |
| | Au suspension in PMMA monomer | N/A | Spherical | None | Mechanical properties affecting flexural strength and elastic modulus, thermal conductivity, density and hardness | Incorporation of AuNPs into heat-polymerized PMMA resin led to a decrease in the flexural strength and elastic modulus. At the same time, the density, thermal conductivity and hardness increased. | [81] |
| Polyurethane resins (PU, dental aligner; Invisalign) | Solution immersion | <4 nm of AuDAPTs | N/A | 4,6-diamino-2-pyrimidinethiol (DAPT) | Antibacterial | AuDAPT-coated aligners have antibacterial effects on a suspension of *P. gingivalis*, a drug-resistant pathogenic oral bacterium. The strength of these effects depends on the amount of AuDAPT and bacterial concentration. The neighboring area of the material was also affected. AuDAPT-coated aligners slowed biofilm formation and showed favorable biocompatibility. | [82] |

**Table 3.** *Cont.*

| Material | Modification Technique on Materials | Modified Factors of AuNPs | | | Enhanced Activities | Findings | Reference |
|---|---|---|---|---|---|---|---|
| | | Size (nm) | Shape | Surface Modification | | | |
| Polyacrylic acid (PAA) and polyacrylamide (PAm) | Au suspension in AA or Am monomer | AuCl—57.2 to 69.4 nm AuAc—84.2 to 134.3 nm | AuCl—Spherical AuAc—Ellipsoidal | None | Mechanical properties | Polyacrylate-AuNP composites exhibited lower compressive strength compared to the control samples, while their toughness increased. | [83] |
| Resin composites | Filler composition | 12–15 | Spherical | None | Polymerization ratio | AuNPs showed plasmon-enhanced polymerization for green-light-photopolymerizable dental resin with a maximum value at 0.0208 wt%. | [29] |
| | Filler composition | AuCANPs = 14.7 ± 4.7 nm AuCPNPs = 18.7 ± 9.7 nm | N/A | Citrate (AuCAPNPs) Carboxyphenyl (AuCPNPs) | MMP inhibition and mechanical properties | AuNPs are attractive candidates to inhibit MMPs and improve the mechanical properties of resins without cytotoxic/genotoxic effects on cells, and therefore they should be suitable for applications in adhesive resin systems. | [84] |
| | Filler composition | 5 | Spherical | None | Degree of conversion and mechanical property affecting diametral tensile stress | The optimal conditions were a 0.0208 wt% AuNP concentration and 1.4 mW/cm$^2$ light intensity, at which the DTS and DC data were all maximal. | [85] |
| Stainless steels | Solution immersion | 3.5–5.5 | Nanocore | polyoxoborate matrix | Antibacterial | The use of a nanocomposite consisting of AuNPs incorporated into a polyoxoborate matrix (BOA) had been found to effectively decrease the colonization of orthodontic appliances by *S. mutans*. | [86] |

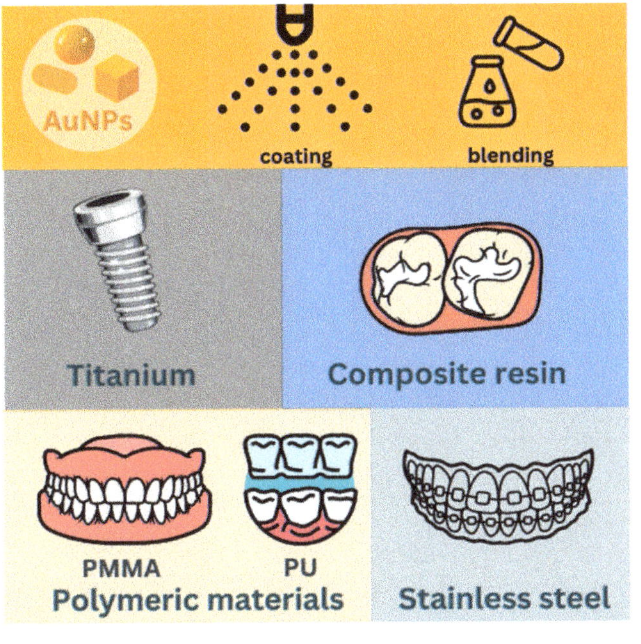


**Figure 3.** AuNP-modified dental materials.

3.2.1. Blending AuNPs in Dental Biomaterials

Polymeric materials are frequently considered potential biomaterials for application in the field of dentistry. Suspensions of AuNPs are incorporated as polymeric monomers or fillers, mainly PMMA and resin composites. PMMA/AuNPs are smoother and more hydrophilic than a pure PMMA surface [27]. Thus, they showed a significant antibiofilm effect in *C. albicans*, *S. mitis*, *S. aureus* and *E. coli* [20]. The presence of PMMA-AuNPs at concentrations above 200 ppm resulted in a considerable reduction (56.67%) in viable fungal cell adhesion in $Au^0$. No statistically significant differences were observed between the 200 and 400 ppm loads. The addition of AuNPs may change the way the surface interacts with its surroundings or make it more negatively charged. PMMA-AuNPs exerted a predominantly repulsive effect on the negatively charged fungal cell wall. This procedure was considered essential for the antiadhesion effect [87]. AuNPs are promising candidates for suppressing MMPs by chelating $Zn^{2+}$ bound in the active sites of MMPs and enhancing the mechanical characteristics of resins, without any harmful effects on cells. Consequently, they are well suited for use in adhesive resin systems [84]. AuNPs, as fillers, exhibited plasmon-enhanced polymerization for green light photopolymerization of resin composites [29,85].

3.2.2. Coating AuNPs in Dental Biomaterials

Titanium is the most commonly used material in implant dentistry due to its biocompatibility and mechanical properties. Titanium must be treated to induce polarization of the surface, which can attract osteogenic cells to promote bone regeneration [15,22]. In addition, AuNPs can promote osteoblast differentiation and be strongly immobilized on the titanium surface via Au–S bonding [14,22]. Titanium dioxide nanotubes or titania nanotubes (TNTs) have attractive properties in terms of cytocompatibility, osseointegration and photocatalytic activity. TNTs were prepared from titanium surfaces by electrochemical anodization in a 0.5–1.0 wt % HF solution [73,74,88]. Then, AuNPs were deposited on TNTs' surface by soaking under UV light illumination [74]. In addition, 3-aminopropyltrimethoxysilane (3-APS) was used as a silane coupling agent for the immobilization of AuNPs on the TNT [88]. Xu et al. found that electrodeposition is an effective method for immobilizing AuNPs on the titanium surface by tuning AuNPs into the titania nanotubes to improve osteogenesis by modulating macrophages' polarization [15]. Magnetron sputtering and ion plasma sputtering are other methods for coating AuNPs on TNT [73]. Chitosan-AuNPs are conjugated with growth factors and DNA to stimulate bone formation and induce osseointegration of titanium dental implants [75]. Furthermore, AuNPs' deposition on titania nanotubes causes antibiofilm enhancement and promotes soft tissue healing [74]. The multi-coating of AuPt-NPs on titania nanotubes under visible-light irradiation up to 600 nm plays a crucial role in the antibacterial effect and filopodia behavior of hMSCs, which promoted osteogenic functionality in hMSCs [6]. Moreover, an AuDAPT-coated dental aligner (polyurethane resins) showed not only impressive antibacterial activity against *P. gingivalis* but also favorable prevention of biofilm formation [82]. Utilizing a nanocomposite composed of AuNPs integrated into a polyoxoborate matrix (BOA) has been shown to significantly reduce the colonization of stainless-steel orthodontic appliances by *S. mutans* [86].

3.2.3. Effect of AuNPs on the Mechanical Properties of Dental Biomaterials

AuNPs exhibit promising potential for enhancing the mechanical properties of resins used in adhesive systems with satisfactory biocompatibility. Additionally, these combined properties make AuNPs a valuable component for optimizing adhesive resin systems in various applications. The incorporation of AuNPs into PMMA led to a decrease in tensile strength, flexural strength and elastic modulus [27,81]; however, other mechanical properties, such as density, thermal conductivity and microhardness, increased [20,81]. In contrast, Oyar, Sana, Nasseri and Durkan [61] found that the addition of AuNPs to heat-cured PMMA at lower concentrations (0.05%) significantly increased the flexural strength compared with higher concentrations (0.20%). The compressive strength of the polyacrylate–AuNP composites was greater than that of the control samples, although their toughness was improved [83]. However, the difference in the size of AuNPs did not affect the flexural strength of the PMMA. Even when the highest dose of AuNPs (400 ppm) was added, the PMMA–AuNPs still showed a thermogram comparable to that of the control during the deterioration process. The presence of AuNPs did not affect the glass transition phase of the denture acrylic. Although adding AuNPs to denture acrylic provides an antibacterial benefit, there was a considerable increase in ΔE values as the dose of AuNPs was increased from 100 to 400 ppm [87].

### 3.3. *AuNPs and Their Related Host Cells in the Oral Cavity*

AuNPs are widely used in the form of coatings on many materials to enhance not only steps in the healing process such as cell adhesion, proliferation and differentiation but also antibacterial, antioxidant and drug-molecule-carrying activities. Gold and its alloys have been used in dental applications and reconstructive surgery. Due to their biocompatibility, durability and inertness, they do not cause an immune response or allergic reactions in the body. AuNPs can conjugate various drug molecules and have specific functions in targeted therapy. The optimal concentration, size and shape of AuNPs for use in nanomaterials are vital factors in stem cell therapies, as well as tissue engineering and regenerative medicine [65,66].

#### 3.3.1. Osteogenic Potential of Stem Cells

The alveolar bone is one of the important hard tissues in the oral cavity that surrounds the teeth. Bone adaptation plays a major role in the healing process of alveolar bone defects that generally result from periodontitis, trauma and infection [89]. Osseointegration plays a major role in successful implant anchorage. Crucial factors that influence the success rates of dental implant placement in low-density bone are implant material, implant design and surgical procedures [90]. Recently, researchers have focused on bone tissue engineering applications utilized by AuNPs [14,22,26]. The mechanism by which AuNPs induce bone repair remains unknown. Prior research has demonstrated that conjugated AuNPs have significant potential for reducing the viability of osteoclasts while also promoting osteogenic development without causing any harm. These nanoparticles possess favorable attributes for the treatment of osteoporotic disorders [30]. A titanium implant surface immobilized by an AuNP layer induced the mRNA expression of osteogenic differentiation-specific biomarkers Runx2, OCN, Col-1, OPN and BMP 2, which resulted in osseous implant interface formation [14,56,57,79,91]. The expression of Runx2 genes in the nucleus is the key osteogenic transcription factor and is upregulated by the *p38/MAPK*, *ERK/MAPK* or *Wnt/β-catenin* signaling pathways, which can be mechanically activated by AuNPs [92]. Moreover, AuNPs regulated the activation of Yes-associated protein (YAP), resulting in osteogenic differentiation [4,56]. AuNPs demonstrated a higher percentage of cell survival in human MG-63 cell lines compared to the control group, indicating their osteoinductive potential [57]. AuNPs may activate cellular autophagy, which affects the cytoskeletal structure to promote osteogenic differentiation [92,93]. In addition, AuNPs inhibit adipogenic differentiation [3,14,22,58,79]. AuNPs promoted the osteogenic functionality of hMSCs under 600 nm visible-light irradiation because of the combined effect of photothermal scattering and visible-light low-level laser therapy (LLLT) [6,94]. Moreover, AuNPs immobilized on the titanium surface led to increased ADSC or MSC proliferation and a maximized level of ALP activity [14,22,79]. The modification of nanostructures affected the viability and differentiation in rat bone marrow MSCs [5]. Xu, He, Zhang, Ma, Zhang and Song [15] illustrated that AuNPs lowered inflammatory responses by reducing microvessel density and mediated the M1–M2 transition of macrophage polarization, which has the potential to promote macrophage-mediated osteogenesis. AuNPs enhanced osteogenic differentiation in a size-dependent manner [4,95]. Smaller AuNPs (25–35 nm) entrapped within the anodized pure titanium surface can promote M2-polarized macrophages and enhance osteogenesis [15]. Furthermore, 30 and 50 nm AuNPs appeared to promote the highest osteogenic differentiation of ADSCs and hMSCs [3,4]. AuNPs of 45 nm could promote osteogenic differentiation of PDL stem cells and bone regeneration more than 13 nm AuNPs through autophagy [93]. Ch-AuNPs/c-myb suppressed osteoclastogenesis and promoted osteogenesis [75]. All sizes and shapes of AuNPs promoted the differentiation of various stem cells toward osteoblasts, but 30–50 nm AuNPs demonstrated an extraordinary capacity for bone regeneration [22]. In many reports, AuNPs of various shapes such as nanospheres, nanostars and nanorods affected osteogenic differentiation [4]. Li, Li, Zhang, Wang, Kawazoe and Chen [4] also found that 40 nm AuNP nanorods inhibited the osteogenesis ability of hMSCs. AuNPs can exhibit not only osteogenic differentiation but also extracellular matrix mineralization [92]. In another study, surface modification did not affect osteogenic differentiation and stimulated osteogenesis by increasing extracellular matrix mineralization. AuNP-COOH showed upregulation of proliferation growth factors of hMSCs (FGF-2 and TGF-β); however, it showed a reduction in ALP activity and matrix mineralization in hMSCs [7].

#### 3.3.2. Gingival Fibroblast and Epithelial Cells

The oral epithelium is the outermost barrier against the oral environment and has contact with vascularized connective tissue. AuNPs can accelerate wound healing by promoting cell migration, angiogenesis and collagen deposition [68]. Moreover, AuNPs/TiO2-NTs also promote fibroblast adhesion, proliferation and migration by upregulating the gene expression levels of fibronectin and pFAK without any irradiation [74]. AuNPs have procoagulant activity, which gives them superior adhesion ability to promote the accumulation of platelets and accelerate the coagulation process. AuNPs can accelerate wound healing due to having denser tissue than new collagen on the CCAu3 group, which illustrates the antioxidant effects and relates to the corresponding gene expression [68]. With the underlying connective tissue, fibroblasts proliferate and migrate into the wound bed and deposit new extracellular matrix, followed by a complex cascade of intracellular biological pathways [68,74]. Thus, the oral epithelium has numerous factors associated with the healing process, for instance, cells, cytokines, saliva and microorganisms.

#### 3.3.3. Dental Pulp Stem Cells (DPSCs)

The biocompatibility and proliferation rate of DPSCs produced from SHED were shown to be enhanced by spherical AuNPs [65]. The incorporation of a novel calcium phosphate cement made of AuNPs enhanced the performance of human DPSCs. The observed improvement was characterized by an increase in cell adhesion and proliferation, as well as an enhancement in osteogenic differentiation [56]. In situ SERS revealed that mitochondrial nanoprobes (MT-AuNPs/MLS@RGD-PEG-AuNPs) enhanced the accelerated differentiation of DPSCs by increasing the MMP to stimulate more ATP production via the thermoplasmonic effect of AuNPs. The results indicated an increase in the expression of glucose and hydroxyproline within DPSCs during the cell differentiation process [55].

### *3.4. The Possible Use of AuNPs in Clinical Dental Applications*

AuNPs have demonstrated significant potential in several clinical dental applications because of their distinctive features, including their compatibility with living organisms, simplicity of modification and optical qualities.

#### 3.4.1. Diagnostic Imaging and Oral Cancer Detection

AuNPs exhibit unique optical properties, including strong absorption and scattering of light due to SPR, which can enhance the contrast in imaging techniques such as X-rays, computed tomography (CT) and magnetic resonance imaging (MRI). Their ability to scatter light and absorb radiation improves the visibility of dental structures and other abnormalities. This phenomenon arises from the collective oscillation of free conduction electrons in the metal when stimulated by incident electromagnetic radiation, which causes the visible color to change [50]. The wavelength at which SPR occurs is determined by several factors, such as the dimensions, morphology and dielectric surroundings of the nanoparticles [63,96]. By controlling these parameters, the SPR band of AuNPs can be specifically tuned across a wide range of wavelengths, from visible to near-infrared (NIR) regions. AuNPs are used as bioimaging applications, as fluorescent probes and X-ray contrast agents at very low concentrations [1,63]. AuNPs have also been employed in biosensors for an early detection of oral cancer. These sensors can detect specific biomarkers associated with cancer, providing a non-invasive and sensitive method for early diagnosis. Functionalized AuNPs can be designed to bind to specific biomarkers or cells, allowing for targeted imaging of particular tissues or pathogens in the oral cavity. In addition, AuNPs can bind many proteins and drugs and can be actively targeted to cancer cells by overexpressing a cell's biomarkers [97]. Moreover, due to cellular uptake, imaging can localize the intracytoplasmic region of the cell [96]. The AuNP–PLL complex monitored the cellular behavior of DPSCs using micro-CT [59].

3.4.2. Therapeutic Applications

AuNPs act as drug carriers conjugated by several biomolecules [62,96,98–100]. Functionalization of AuNPs with biomolecules (e.g., antibodies, peptides, DNA) allows for targeted delivery and specific interactions with biological targets. This can be exploited in receptor-mediated tumor cell targeting, enhanced biocompatibility and biorecognition. AuNPs have been used to modify dental materials to impart antibacterial properties that can prevent the formation of biofilms that adhere to surfaces and are often resistant to conventional antibiotics. Researchers have developed materials that can inhibit bacterial growth and reduce the risk of infections by incorporating AuNPs into dental composites or coatings. Therefore, AuNPs are versatile nanocarriers for targeted drug delivery systems in cancer therapy. They have been shown to play crucial roles in cytotoxicity, protection from enzymatic degradation and enhanced cellular uptake [98,101,102]. AuNPs have strong cationic attractions to negatively charged surfaces, so can adhere on the microbial membrane through electrostatic interaction [39,48,103]. Moreover, AuNPs strongly electrostatically adsorbed with lysine, which is the most abundant amino acid on the membrane of Gram-positive bacteria. This caused various distortions in the structure of microbes, such as changes in permeability, osmolarity, electron transport to reach permanent pores and the subsequent loss of cellular content, which causes cell death [12,38,45,69]. On the other hand, some studies found that the thicker peptidoglycan cell walls of Gram-positive bacteria may protect their cells from rupturing. Therefore, AuNPs are more effective against Gram-negative bacteria [20,45]. The strong electrostatic attraction between AuNPs and the bacterial cell wall interrupted adhesin-mediated interaction, which inhibits pathogenic biofilm formation [103]. At greater concentrations, AuNPs were as effective an antibacterial agent against *S. oralis* as chlorhexidine [44]. In addition, a relatively low concentration of AuNPs is able to trigger apoptosis in malignant cells and causes lower cytotoxicity [16,64,96]. AuNPs can easily identify tumor cells or pathogenic microbial cells, due to specific environmental factors such as an acidic pH or hypoxic conditions [18,19,39,98,104]. AuNPs cause a drastic induction of cell cycle arrest and DNA damage [72]. Chitosan–AuNP–curcumin composites had pH-responsive controlled release behaviors that led to significantly improved drug delivery systems under localized acidic conditions [39]. Lysozyme-functionalized gold nanoclusters (AuNCs) can attach the N-acetylglucosamine on the bacterial cell walls through the lysozyme interaction. This binding triggered the detection of bacteria, leading to drug release and subsequent antibacterial activity. [37].

A photothermal effect is induced by AuNPs under NIR irradiation, which is commonly used in non-invasive cancer therapy. This can produce heat energy from absorption and scattering light to stimulate cellular apoptosis and denature proteins. This process can be used in photothermal therapy to selectively destroy harmful cells or bacteria in the mouth, such as those associated with oral cancer or periodontitis, when temperatures reach more than 50 °C [19,68,69,105]. However, further research should examine the comparative impact on bacteria in biofilms. However, it is unlikely that the formation of biofilms alone would provide sufficient protection for bacteria against extremely elevated localized temperatures (73 °C) [100]. AuNPs bind various molecules that can be functionalized as targeted therapy by a specific biomarker and act as light-responsive nanomaterials [106]. In addition, AuNPs eliminate biofilms by thermal degradation under NIR light and may be effective alternatives to antibiotics in the treatment of bacterial infections [39,106]. *S. aureus* and *E. coli* can be eliminated by functionalized AuNPs under NIR light, as well as methicillin-resistant *S. aureus* (MRSA) [39,105]. PTT is an effective antibacterial mechanism that does not lead to antibiotic resistance [68]. In summary, an increased concentration of AuNPs results in a greater absorption rate of photon energy from the laser. This energy can then be converted into heat, leading to an elevation in cell temperature. Consequently, the viability of cells, particularly cancer cells, which are sensitive to heat, is reduced [54].

AuNPs acting as a photosensitizer absorb photon energy and convert the surrounding oxygen molecules into highly toxic ROS. This is called the photocatalytic process [19,39,74]. Routine PDT is activated by a common photosensitizer (i.e., methylene blue) for eliminating bacterial biofilm. AuNP-conjugated methylene blue was an attractive photosensitizer for efficient ROS generation, including singlet oxygen ($^1O_2$) or superoxide ($O_2^{\cdot-}$) through resonance energy transfer (RET) from AuNCs to methylene blue. This illustrated similar bacterial growth inhibition and bactericidal effect [107]. AuNP-conjugated methylene blue capped with bovine serum albumin showed significant antimicrobial PDT against *S. mutans* under white-light LED irradiation for approximately 1 min. The AuNC–MB combination did not exhibit any antibacterial activity in the absence of LED light [49]. Conversely, the electrons could be stored in AuNPs and released into the environment without light irradiation. Consequently, the electrons interacted with oxygen to produce ROS, which resulted in the disruption of the bacterial membrane [47,73,74]. Siberian ginseng-AuNPs had increased ROS generation and mitochondrial membrane permeability of B16 cells, leading to the release of proapoptotic proteins [53]. AuNPs directly enter the bacterial cell and can interrupt the ATP synthesis and whole-cell mechanisms, including membrane damage and cellular uptake [38,69]. Conversely, AuNPs exhibit either no antibacterial activity or very weak antibacterial properties [62,108–110]. After the Turkevich method, the most common AuNPs preparation method involves the chemical reduction of $Au^{3+}$ salts to $Au^0$, which induces ROS formation and can alter macromolecules, resulting in effective antimicrobial activity against a wide variety of microorganisms [38,39,109]. The deposition of gold ions on different cells showed that metal ions can damage DNA, inhibit replication, disturb cellular or membrane proteins and decrease the production of ATP [46]. The concurrent utilization of low-temperature plasma, ROS generation and AuNPs caused significant disruption of cells, resulting in the release of internal components from several cells [48]. LLLT at 470 and 600 nm played an important role in the antibacterial performance of AuPt/PtAu $TiO_2$ [6]. Cells irradiated with AuNPs exhibited a notable decrease in MCF-7 cancer cell density in comparison to cells treated with AuNPs and laser separately [54]. Therefore, ROS generation in an aerobic environment can cause the oxidation of protein and nucleic acids, causing cellular damage [38,69,74]. It also led to impairment of mitochondria function, ultimately resulting in cell death [66]. AuNPs decorated on titanium oxide nanotubes showed a photocatalytic memory effect due to their unique heterogeneous structures and were activated not only via NIR and visible-light irradiation according to the size and shape of AuNPs, but also via a simple ultrasound treatment [6,74,111]. ROS formation occurs in an aerobic environment; the anaerobic Gram-positive bacteria *S. mitis* may not be affected by AuNPs [20]. This hypoxic condition occurs when cells do not receive enough oxygen to induce DNA damage caused by free radicals containing oxygen, and it protects the cells from radiation [54].

#### 3.4.3. Restorative Dentistry

AuNPs are being explored for use in dental restorations to improve the mechanical properties and longevity of materials such as PMMA, dental resins and cements. AuNPs enhanced the mechanical characteristics of resins without causing any harmful effects on cells, and therefore should be suitable for applications in adhesive resin systems [84]. Moreover, AuNPs display SPR upon irradiation at specific frequencies, enabling them to improve the polymerization ratio through the plasmonic effect in resin composites and facilitating polymer chain formation around the metallic nanoparticles, which increased the degree of conversion [29,85]. Higher concentrations of AuNPs in PMMA/AuNPs composites resulted in a modest increase in Vicker's microhardness [20]. The density and thermal conductivity of PMMA were also increased. However, AuNPs decreased the tensile strength, flexural strength and elastic modulus of PMMA [27,81]. The flexural strength values of NPs doped with 400 ppm revealed a substantial increase compared to dosages of 0–200 ppm. The incorporation of NPs might enhance the mechanical strength of denture base structures without adversely affecting the original PMMA material [87]. On the other hand, nanoparticles had the potential to form vast clusters due to concentrations of stress at the sites of agglomeration, which resulted in a considerable attenuation of flexural strength at higher concentrations (0.20%) compared to lower concentrations (0.05%) of the same particle size. Thus, the modification of AuNPs at lower concentrations did not affect the mechanical properties of the materials but enhanced the flexural strength of PMMA. Hence, the addition of AuNPs of varying sizes to PMMA did not have a substantial impact on its flexural strength [61]. Thus, AuNPs can be used for biomedical purposes without decreasing the mechanical properties of the materials.

#### 3.4.4. Regenerative Dentistry

AuNPs can be used to enhance the growth and differentiation of stem cells, potentially aiding in the repair or regeneration of dental tissue (as reviewed in Section 3.3). AuNPs exhibited an interaction with free radicals, resulting in a decrease in DPPH [52]. Spherical AuNPs exhibited the greatest inhibitory activity on the DPPH radical, while polyhedral AuNPs did not show antioxidant properties. AuNPs synthesized from plant extracts also exhibited better antioxidant activity [41,50]. The antioxidant activity of AuNPs was found to be dose-dependent and was shown to be similar to the ability of ascorbic acid (standard) to scavenge DPPH. At a concentration of 50 μg/mL, CuAuNp demonstrated a maximum DPPH scavenging performance of 90.3% [45]. Green AuNPs using the microwave irradiation synthesis method and *Saussurea obvallata* plant extract showed improved antioxidative action at a concentration of 1000 μg/mL [52]. Moreover, heat production from the oscillation of ursodeoxycholic-AuNPs (UDCA-AuNPs) under NIR irradiation can inhibit proinflammatory cytokines [51]. MLS@RGD-PEG-AuNPs accelerated the differentiation of DPSCs through the thermoplasmonic effect of AuNPs by in situ SERS, which effectively regulated mitochondrial metabolism [55]. Green synthesis of red sandal AuNPs inhibited protein denaturation to a degree comparable to commercial painkillers by reducing chemotaxis and the endothelial leukocyte contact [50]. As the concentration of the extract increased, there was a corresponding increase in anti-inflammatory activity that was similar to the conventional medication standard. CuAuNP shown 79.6% greater anti-inflammatory activity at a concentration of 50 μg/mL compared to diclofenac [45].

### *3.5. The Possible Toxicity of AuNPs in the Dental Field*

AuNPs demonstrate excellent compatibility with mammalian cells at lower concentrations [42]. The cytotoxicity of AuNPs is correlated with the surface area of the AuNPs because of the size and concentration of particles taken up into the cells. Cellular oxidative stress leads to an elevation in lactate dehydrogenase (LDH) release, the initiation of apoptosis and the formation of intracellular ROS [66]. AuNPs can be regarded as more biocompatible than AgNPs due to the absence of ROS production in their mechanism of action In contrast, ultra-small particles (1–2 nm) tend to be more toxic, not only due to ROS generation but also because their small size can cause irreversible binding to biopolymers [112]. The size of AuNPs affects the distribution of particles to different organs: larger AuNPs are taken up by macrophages and have somewhat restricted distribution while being more widely distributed throughout the systems, particularly the liver and spleen. Particle size and protein adsorption, together with the capacity of certain cells to internalize AuNPs by endocytosis, determine the method of uptake. AuNPs can induce harmful effects on several systems [95]. Previous studies have demonstrated that nanospheres and rods have higher toxicity levels in comparison to star-shaped, flower-shaped and prismatic AuNPs [112]. Moreover, rod-shaped AuNPs caused more toxicity in SHED than spherical-shaped ones [65]. As a result, AuNPs should be carefully chosen in terms of their size, concentration and duration of clinical application [17,65,96].

### 3.6. Future Perspectives on the Use of AuNPs in the Dental Field

AuNPs demonstrate certain essential characteristics resulting from their biological, mechanical and optical properties. Recent developments in dental technology emphasize the use of innovative materials for therapeutic and regenerative purposes. AuNPs offer enormous potential for tissue engineering, specifically in the processes of stem cell proliferation and differentiation, and exhibit excellent biocompatibility. Thus, both normal and compromised conditions could be improved by utilizing AuNPs in dental treatment. Future developments in the dental field involving AuNPs are expected to encompass their use in diagnostic and imaging technologies [63,96], dental implantology [14,22,57,69], pulp tissue regeneration [59] and periodontal regeneration [58,113]. AuNPs have the potential to be used in advanced sensing technologies such as oral cancer detection, oral biomarkers and localizing specifically targeted cells or tissue. Several reviews have highlighted the benefits of modifying AuNPs with titanium implants to improve osseointegration [14,22,57,69]. Furthermore, the favorable biological characteristics of the regeneration of bone and tissues could enhance dental therapy, incorporating pulp regeneration, soft tissue grafting, bone augmentation and soft tissue healing approaches [58,59,113]. The application of AuNPs on implant components, dental appliances, prostheses and aligners has the potential to enhance both the biological response and antibacterial properties. Therefore, the use of AuNPs in dentistry may limit failures and enhance treatment outcomes.

## 4. Conclusions

In conclusion, this review highlights the promising potential of AuNPs in dental materials, particularly through enhancing biological, mechanical and optical properties. Despite the challenges associated with the synthesis and stability of AuNPs, their modifications to titanium, PMMA and resin composites have demonstrated significant benefits, such as improved antibiofilm activity, osteogenesis and mechanical strength. The SPR effect contributes to their application by enhancing resin polymerization. Nevertheless, further research is needed to explore AuNP interactions with other materials, such as zirconia and PEEK, and to develop advanced methods to optimize their use in dental therapies for diverse patient populations.

**Author Contributions:** Conceptualization, S.J. and P.R.; methodology, S.J. and P.R.; software, S.J.; validation, S.J., D.P., A.Y., C.R. and P.R.; formal analysis, S.J. and P.R.; investigation, S.J. and P.R.; resources, S.J. and P.R.; data curation, S.J. and P.R.; writing—original draft preparation, S.J.; writing—review and editing, S.J., D.P., A.Y., C.R. and P.R.; visualization, D.P., A.Y., C.R. and P.R.; supervision, P.R.; project administration, P.R.; funding acquisition, P.R. All authors have read and agreed to the published version of the manuscript.

**Funding:** This research was funded by Chiang Mai University, Chiang Mai, Thailand.

**Data Availability Statement:** Not applicable.

**Conflicts of Interest:** The authors declare no conflicts of interest.

## References


1. AlKahtani, R.N. The implications and applications of nanotechnology in dentistry: A review. *Saudi Dent. J.* **2018**, *30*, 107–116. [CrossRef] [PubMed]
2. Umapathy, V.R.; Natarajan, P.M.; SumathiJones, C.; Swamikannu, B.; Johnson, W.M.S.; Alagarsamy, V.; Milon, A.R. Current trends and future perspectives on dental nanomaterials—An overview of nanotechnology strategies in dentistry. *J. King Saud Univ.-Sci.* **2022**, *34*, 102231. [CrossRef]
3. Ko, W.K.; Heo, D.N.; Moon, H.J.; Lee, S.J.; Bae, M.S.; Lee, J.B.; Sun, I.C.; Jeon, H.B.; Kwon, I.K.; Park, H.K. The Effect of Gold Nanoparticle Size on Osteogenic Differentiation of Adipose-derived Stem Cells. *J. Colloid Interface Sci.* **2015**, *438*, 68. [CrossRef] [PubMed]
4. Li, J.; Li, J.J.; Zhang, J.; Wang, X.; Kawazoe, N.; Chen, G. Gold nanoparticle size and shape influence on osteogenesis of mesenchymal stem cells. *Nanoscale* **2016**, *8*, 7992–8007. [CrossRef]
5. Yuan, L.; Qi, X.; Qin, G.; Liu, Q.; Zhang, F.; Song, Y.; Deng, J. Effects of Gold Nanostructures on Differentiation of Mesenchymal Stem Cells. *Colloids Surf. B Biointerfaces* **2019**, *184*, 110494. [CrossRef] [PubMed]
6. Moon, K.S.; Park, Y.B.; Bae, J.M.; Choi, E.J.; Oh, S.H. Visible Light-Mediated Sustainable Antibacterial Activity and Osteogenic Functionality of Au and Pt Multi-Coated $TiO_2$ Nanotubes. *Materials* **2021**, *14*, 5976. [CrossRef]
7. Li, J.J.; Kawazoe, N.; Chen, G. Gold nanoparticles with different charge and moiety induce differential cell response on mesenchymal stem cell osteogenesis. *Biomaterials* **2015**, *54*, 226–236. [CrossRef] [PubMed]
8. Pissuwan, D.; Poomrattanangoon, S.; Chungchaiyart, P. Trends in Using Gold Nanoparticles for Inducing Cell Differentiation: A Review. *ACS Appl. Nano Mater.* **2022**, *5*, 3110–3120. [CrossRef]
9. Wang, Y.; Quinsaat, J.E.Q.; Ono, T.; Maeki, M.; Tokeshi, M.; Isono, T.; Tajima, K.; Satoh, T.; Sato, S.I.; Miura, Y.; et al. Enhanced dispersion stability of gold nanoparticles by the physisorption of cyclic poly(ethylene glycol). *Nat. Commun.* **2020**, *11*, 6089. [CrossRef] [PubMed]
10. Tukova, A.; Nie, Y.; Yaraki, M.T.; Tran, N.; Wang, J.; Rodger, A.; Gu, Y.; Wang, Y. Shape dependent protein-induced stabilization of gold nanoparticles: From a protein corona perspective. *Aggregate* **2023**, *4*, e323. [CrossRef]
11. Engstrom, A.M.; Faase, R.A.; Marquart, G.W.; Baio, J.E.; Mackiewicz, M.R.; Harper, S.L. Size-Dependent Interactions of Lipid-Coated Gold Nanoparticles: Developing a Better Mechanistic Understanding through Model Cell Membranes and in vivo Toxicity. *Int. J. Nanomed.* **2020**, *15*, 4091–4104. [CrossRef]
12. Kiarashi, M.; Mahamed, P.; Ghotbi, N.; Tadayonfard, A.; Nasiri, K.; Kazemi, P.; Badkoobeh, A.; Yasamineh, S.; Joudaki, A. Spotlight on therapeutic efficiency of green synthesis metals and their oxide nanoparticles in periodontitis. *J. Nanobiotechnol.* **2024**, *22*, 21. [CrossRef] [PubMed]
13. Nandhini, J.; Karthikeyan, E.; Rajeshkumar, S. Nanomaterials for wound healing: Current status and futuristic frontier. *Biomed. Technol.* **2024**, *6*, 26–45. [CrossRef]
14. Ko, W.K.; Kim, S.J.; Heo, D.N.; Han, I.B.; Kim, S.; Kwon, I.K.; Sohn, S. Double layers of gold nanoparticles immobilized titanium implants improve the osseointegration in rabbit models. *Nanomed. Nanotechnol. Biol. Med.* **2020**, *24*, 102129. [CrossRef] [PubMed]
15. Xu, B.; He, Y.; Zhang, Y.; Ma, Z.; Zhang, Y.; Song, W. In Situ Growth of Tunable Gold Nanoparticles by Titania Nanotubes Templated Electrodeposition for Improving Osteogenesis through Modulating Macrophages Polarization. *ACS Appl. Mater. Interfaces* **2022**, *14*, 50520–50533. [CrossRef] [PubMed]
16. Moloudi, K.; Khani, A.; Najafi, M.; Azmoonfar, R.; Azizi, M.; Nekounam, H.; Sobhani, M.; Laurent, S.; Samadian, H. Critical parameters to translate gold nanoparticles as radiosensitizing agents into the clinic. *Wiley Interdiscip. Rev. Nanomed. Nanobiotechnol.* **2023**, *15*, e1886. [CrossRef]
17. Mironava, T.; Hadjiargyrou, M.; Simon, M.; Rafailovich, M.H. Gold nanoparticles cellular toxicity and recovery: Adipose Derived Stromal cells. *Nanotoxicology* **2014**, *8*, 189–201. [CrossRef] [PubMed]
18. Mahhengam, N.; Kazemnezhad, K.; Budi, H.S.; Ansari, M.J.; Bokov, D.O.; Suksatan, W.; Thangavelu, L.; Siahmansouri, H. Targeted therapy of tumour microenvironment by gold nanoparticles as a new therapeutic approach. *J. Drug Target.* **2022**, *30*, 494–510. [CrossRef]
19. Devi, R.S.; Girigoswami, A.; Siddharth, M.; Girigoswami, K. Applications of Gold and Silver Nanoparticles in Theranostics. *Appl. Biochem. Biotechnol.* **2022**, *194*, 4187–4219. [CrossRef]
20. Ivanovic, V.; Popovic, D.; Petrovic, S.; Rudolf, R.; Majerič, P.; Lazarevic, M.; Djordjevic, I.; Lazic, V.; Radunovic, M. Unraveling the Antibiofilm Activity of a New Nanogold Resin for Dentures and Epithesis. *Pharmaceutics* **2022**, *14*, 1513. [CrossRef]
21. Ferreira-Gonçalves, T.; Ferreira, D.; Ferreira, H.A.; Reis, C.P. Nanogold-based materials in medicine: From their origins to their future. *Nanomedicine* **2021**, *16*, 2695–2723. [CrossRef] [PubMed]
22. Heo, D.N.; Ko, W.K.; Lee, H.R.; Lee, S.J.; Lee, D.; Um, S.H.; Lee, J.H.; Woo, Y.H.; Zhang, L.G.; Lee, D.W.; et al. Titanium dental implants surface-immobilized with gold nanoparticles as osteoinductive agents for rapid osseointegration. *J. Colloid Interface Sci.* **2016**, *469*, 129–137. [CrossRef] [PubMed]
23. Taha, E.Y.; Elmahdy, M.M.B.; Masry, S.; Elsayed, M.E. Effect of nanogold particles addition on dimensional stability of complete denture base material: An in-vitro study. *BMC Oral Health* **2023**, *23*, 153. [CrossRef] [PubMed]
24. Darvish, S.; Budala, D.-G.; Goriuc, A. Antibacterial Properties of an Experimental Dental Resin Loaded with Gold Nanoshells for Photothermal Therapy Applications. *J. Funct. Biomater.* **2024**, *15*, 100. [CrossRef] [PubMed]

25. Madeira, S.; Barbosa, A.; Moura, C.G.; Buciumeanu, M.; Silva, F.S.; Carvalho, O. Aunps and Agµps-functionalized zirconia surfaces by hybrid laser technology for dental implants. *Ceram. Int.* **2020**, *46*, 7109–7121. [CrossRef]
26. Tuikampee, S.; Chaijareenont, P.; Rungsiyakull, P.; Yavirach, A. Titanium Surface Modification Techniques to Enhance Osteoblasts and Bone Formation for Dental Implants: A Narrative Review on Current Advances. *Metals* **2024**, *14*, 515. [CrossRef]
27. Marić, I.; Zore, A.; Rojko, F.; Škapin, A.S.; Štukelj, R.; Učakar, A.; Vidrih, R.; Veselinović, V.; Gotić, M.; Bohinc, K. Antifungal Effect of Polymethyl Methacrylate Resin Base with Embedded Au Nanoparticles. *Nanomaterials* **2023**, *13*, 2128. [CrossRef] [PubMed]
28. Ibrahim, S.H.; Eisa, M.S.M.S. Microshear Bond Strength, Ultramorphological, and Elemental Assessment of Gold–Silver Nanoparticle-treated Dentin Bonded to Resin Composite with Different Adhesive Modes. *J. Contemp. Dent. Pract.* **2022**, *23*, 679–687. [CrossRef] [PubMed]
29. Szalóki, M.; Csarnovics, I.; Bonyár, A.; Ungor, D.; Csapó, E.; Sápi, A.; Hegedűs, C. Plasmonic Effect of Gold-Patchy Silica Nanoparticles on Green Light-Photopolymerizable Dental Resin. *Nanomaterials* **2023**, *13*, 2554. [CrossRef]
30. Higino, T.; França, R. Drug-delivery nanoparticles for bone-tissue and dental applications. *Biomed. Phys. Eng. Express* **2022**, *8*, 042001. [CrossRef]
31. Mobed, A.; Hasanzadeh, M.; Seidi, F. Anti-bacterial activity of gold nanocomposites as a new nanomaterial weapon to combat photogenic agents: Recent advances and challenges. *RSC Adv.* **2021**, *11*, 34688–34698. [CrossRef] [PubMed]
32. Malik, M.A.; Hashmi, A.A.; Al-Bogami, A.S.; Wani, M.Y. Harnessing the power of gold: Advancements in anticancer gold complexes and their functionalized nanoparticles. *J. Mater. Chem. B* **2024**, *12*, 552–576. [CrossRef]
33. Riley, R.S.; Day, E.S. Gold nanoparticle-mediated photothermal therapy: Applications and opportunities for multimodal cancer treatment. *Wiley Interdiscip. Rev. Nanomed. Nanobiotechnol.* **2017**, *9*, e1449. [CrossRef] [PubMed]
34. Oliveira, B.B.; Ferreira, D.; Fernandes, A.R.; Baptista, P.V. Engineering gold nanoparticles for molecular diagnostics and biosensing. *Wiley Interdiscip. Rev. Nanomed. Nanobiotechnol.* **2023**, *15*, e1836. [CrossRef]
35. Pissuwan, D.; Gazzana, C.; Mongkolsuk, S.; Cortie, M.B. Single and multiple detections of foodborne pathogens by gold nanoparticle assays. *Wiley Interdiscip. Rev. Nanomed. Nanobiotechnol.* **2020**, *12*, e1584. [CrossRef] [PubMed]
36. Milan, J.; Niemczyk, K.; Kus-Liśkiewicz, M. Treasure on the Earth—Gold Nanoparticles and Their Biomedical Applications. *Materials* **2022**, *15*, 3355. [CrossRef]
37. Alsaiari, S.K.; Hammami, M.A.; Croissant, J.G.; Omar, H.W.; Neelakanda, P.; Yapici, T.; Peinemann, K.V.; Khashab, N.M. Colloidal Gold Nanoclusters Spiked Silica Fillers in Mixed Matrix Coatings: Simultaneous Detection and Inhibition of Healthcare-Associated Infections. *Adv. Healthc. Mater.* **2017**, *6*, 1601135. [CrossRef]
38. Niemirowicz-Laskowska, K.; Mystkowska, J.; Łysik, D.; Chmielewska, S.; Tokajuk, G.; Misztalewska-Turkowicz, I.; Wilczewska, A.Z.; Bucki, R. Antimicrobial and Physicochemical Properties of Artificial Saliva Formulations Supplemented with Core-Shell Magnetic Nanoparticles. *Int. J. Mol. Sci.* **2020**, *21*, 1979. [CrossRef]
39. Zhang, Y.; Li, P.; Su, R.; Wen, F.; Jia, Z.; Lv, Y.; Cai, J.; Su, W. Curcumin-loaded multifunctional chitosan gold nanoparticles: An enhanced PDT/PTT dual-modal phototherapeutic and pH-responsive antimicrobial agent. *Photodiagnosis Photodyn. Ther.* **2022**, *39*, 103011. [CrossRef]
40. Al Shayeb, M.A.; Shetty, N.Y.; Al Jadaa, A.; Kuduruthullah, S. Synthesis and characterisation of gold nanoparticles from acacia leaf and to evaluate anticariogenic activity. *Indian. J. Dent. Res.* **2023**, *34*, 196–198. [CrossRef]
41. Gupta, S.; Jangir, O.P.; Sharma, M. The green synthesis, characterization and evaluation of antioxidant and antimicrobial efficacy of silver and gold nanospheres synthesized using wheat bran. *Asian J. Pharm. Clin. Res.* **2016**, *9*, 103–105. [CrossRef]
42. Solanki, L.A.; Sundari, K.K.S.; Rajeshkumar, S. In-vitro cytotoxicity evaluation of green synthesized gold nanoparticles and its indigenous mouthwash. *J. Pure Appl. Microbiol.* **2021**, *15*, 735–742. [CrossRef]
43. Lavaee, F.; Ranjbar, Z.; Modaresi, F.; Keshavarz, F. The Effect of Gold Nano Particles with Different Sizes on Streptococcus Species. *J. Dent.* **2021**, *22*, 235–242. [CrossRef]
44. Al-Fahham, B.M.; Mohamed, R.A.; Al-Talqani, J.M.T.; Fahad, A.H.; Haider, J. Evaluating Antimicrobial Effectiveness of Gold Nanoparticles against Streptococcus oralis. *Int. J. Dent.* **2023**, *2023*, 9935556. [CrossRef]
45. Dharman, S.; Maragathavalli, G.; Shanmugam, R.; Shanmugasundaram, K. Curcumin Mediated Gold Nanoparticles and Analysis of its Antioxidant, Anti-inflammatory, Antimicrobial Activity Against Oral Pathogens. *Pesqui. Bras. Odontopediatria Clin. Integr.* **2023**, *23*, e220068. [CrossRef]
46. Jebali, A.; Hajjar, F.H.E.; Pourdanesh, F.; Hekmatimoghaddam, S.; Kazemi, B.; Masoudi, A.; Daliri, K.; Sedighi, N. Silver and gold nanostructures: Antifungal property of different shapes of these nanostructures on Candida species. *Med. Mycol.* **2014**, *52*, 65–72. [CrossRef] [PubMed]
47. Meeker, D.G.; Wang, T.; Harrington, W.N.; Zharov, V.P.; Johnson, S.A.; Jenkins, S.V.; Oyibo, S.E.; Walker, C.M.; Mills, W.B.; Shirtliff, M.E.; et al. Versatility of targeted antibiotic-loaded gold nanoconstructs for the treatment of biofilm-associated bacterial infections. *Int. J. Hyperth.* **2018**, *34*, 209–219. [CrossRef] [PubMed]
48. Park, S.R.; Lee, H.W.; Hong, J.W.; Lee, H.J.; Kim, J.Y.; Choi, B.B.; Kim, G.C.; Jeon, Y.C. Enhancement of the killing effect of low-temperature plasma on Streptococcus mutans by combined treatment with gold nanoparticles. *J. Nanobiotechnol.* **2014**, *12*, 29. [CrossRef]
49. Yamamoto, M.; Shitomi, K.; Miyata, S.; Miyaji, H.; Aota, H.; Kawasaki, H. Bovine serum albumin-capped gold nanoclusters conjugating with methylene blue for efficient $^1O_2$ generation via energy transfer. *J. Colloid Interface Sci.* **2018**, *510*, 221–227. [CrossRef]

50. Navya, P.D.; Kaarthikeyan, G.; Rajeshkumar, S.; Garapati, B. Assessment of Antioxidant and Anti-inflammatory Properties of Gold Nanoparticles Synthesized Using Pterocarpus Santa- An In Vitro Study. *J. Popul. Ther. Clin. Pharmacol.* **2023**, *30*, 361–367.
51. Ko, W.K.; Lee, S.J.; Kim, S.J.; Han, G.H.; Han, I.B.; Hong, J.B.; Sheen, S.H.; Sohn, S. Direct Injection of Hydrogels Embedding Gold Nanoparticles for Local Therapy after Spinal Cord Injury. *Biomacromolecules* **2021**, *22*, 2887–2901. [CrossRef] [PubMed]
52. Dalavi, P.A.; V, A.J.; Thomas, S.; Prabhu, A.; Anil, S.; Seong, G.H.; Venkatesan, J. Microwave-Assisted Biosynthesized Gold Nanoparticles Using Saussurea obvallata: Biocompatibility and Antioxidant Activity Assessment. *BioNanoScience* **2022**, *12*, 741–751. [CrossRef]
53. Wu, F.; Zhu, J.; Li, G.; Wang, J.; Veeraraghavan, V.P.; Mohan, S.K.; Zhang, Q. Biologically synthesized green gold nanoparticles from Siberian ginseng induce growth-inhibitory effect on melanoma cells (B16). *Artif. Cells Nanomed. Biotechnol.* **2019**, *47*, 3297–3305. [CrossRef]
54. Habit, H.A.H.; Suardi, N.; Mahmud, S.; Mydin, R.B.S.M.N.; Bakhori, S.K.M.; Mzwd, E. Evaluation of synergistic bioinhibitory effect between low-level laser irradiation and gold nanoparticles on MCF-7 cell line. *J. Nanoparticle Res.* **2023**, *25*, 44. [CrossRef]
55. Wang, J.; Qu, X.; Xu, C.; Zhang, Z.; Qi, G.; Jin, Y. Thermoplasmonic Regulation of the Mitochondrial Metabolic State for Promoting Directed Differentiation of Dental Pulp Stem Cells. *Anal. Chem.* **2022**, *94*, 9564–9571. [CrossRef]
56. Xia, Y.; Chen, H.; Zhang, F.; Bao, C.; Weir, M.D.; Reynolds, M.A.; Ma, J.; Gu, N.; Xu, H.H.K. Gold nanoparticles in injectable calcium phosphate cement enhance osteogenic differentiation of human dental pulp stem cells. *Nanomedicine* **2018**, *14*, 35–45. [CrossRef]
57. Jadhav, K.; Hr, R.; Deshpande, S.; Jagwani, S.; Dhamecha, D.; Jalalpure, S.; Subburayan, K.; Baheti, D. Phytosynthesis of gold nanoparticles: Characterization, biocompatibility, and evaluation of its osteoinductive potential for application in implant dentistry. *Mater. Sci. Eng. C Mater. Biol. Appl.* **2018**, *93*, 664–670. [CrossRef] [PubMed]
58. Lee, D.; Heo, D.N.; Nah, H.R.; Lee, S.J.; Ko, W.K.; Lee, J.S.; Moon, H.J.; Bang, J.B.; Hwang, Y.S.; Reis, R.L.; et al. Injectable hydrogel composite containing modified gold nanoparticles: Implication in bone tissue regeneration. *Int. J. Nanomed.* **2018**, *13*, 7019–7031. [CrossRef]
59. Biz, M.T.; Cucco, C.; Cavalcanti, B.N. Incorporation of AuNP-PLL nanocomplexes in DPSC: A new tool for 3D analysis in pulp regeneration. *Clin. Oral Investig.* **2020**, *24*, 1761–1767. [CrossRef] [PubMed]
60. Topala, F.; Nica, L.M.; Boariu, M.; Negrutiu, M.L.; Sinescu, C.; Marinescu, A.; Cirligeriu, L.E.; Stratul, S.I.; Rusu, D.; Chincia, R.; et al. En-face optical coherence tomography analysis of gold and silver nanoparticles in endodontic irrigating solutions: An in vitro study. *Exp. Ther. Med.* **2021**, *22*, 992. [CrossRef]
61. Oyar, P.; Sana, F.A.; Nasseri, B.; Durkan, R. Effect of green gold nanoparticles synthesized with plant on the flexural strength of heat-polymerized acrylic resin. *Niger. J. Clin. Pract.* **2018**, *21*, 1291–1295. [CrossRef] [PubMed]
62. Mitra, P.; Chakraborty, P.K.; Saha, P.; Ray, P.; Basu, S. Antibacterial efficacy of acridine derivatives conjugated with gold nanoparticles. *Int. J. Pharm.* **2014**, *473*, 636–643. [CrossRef]
63. Lopes, T.S.; Alves, G.G.; Pereira, M.R.; Granjeiro, J.M.; Leite, P.E.C. Advances and potential application of gold nanoparticles in nanomedicine. *J. Cell. Biochem.* **2019**, *120*, 16370–16378. [CrossRef]
64. Kesharwani, P.; Ma, R.; Sang, L.; Fatima, M.; Sheikh, A.; Abourehab, M.A.S.; Gupta, N.; Chen, Z.S.; Zhou, Y. Gold nanoparticles and gold nanorods in the landscape of cancer therapy. *Mol. Cancer* **2023**, *22*, 98. [CrossRef] [PubMed]
65. Abuarqoub, D.; Mahmoud, N.; Alshaer, W.; Mohammad, M.; Ibrahim, A.A.; Al-Mrahleh, M.; Alnatour, M.; Alqudah, D.A.; Esawi, E.; Awidi, A. Biological Performance of Primary Dental Pulp Stem Cells Treated with Gold Nanoparticles. *Biomedicines* **2023**, *11*, 2490. [CrossRef]
66. Ibrahim, B.; Akere, T.H.; Chakraborty, S.; Valsami-Jones, E.; Ali-Boucetta, H. Gold Nanoparticles Induced Size Dependent Cytotoxicity on Human Alveolar Adenocarcinoma Cells by Inhibiting the Ubiquitin Proteasome System. *Pharmaceutics* **2023**, *15*, 432. [CrossRef]
67. Villa-García, L.D.; Márquez-Preciado, R.; Ortiz-Magdaleno, M.; Patrón-Soberano, O.A.; Álvarez-Pérez, M.A.; Pozos-Guillén, A.; Sánchez-Vargas, L.O. Antimicrobial effect of gold nanoparticles in the formation of the Staphylococcus aureus biofilm on a polyethylene surface. *Braz. J. Microbiol.* **2021**, *52*, 619–625. [CrossRef] [PubMed]
68. Zheng, L.; Gu, B.; Li, S.; Luo, B.; Wen, Y.; Chen, M.; Li, X.; Zha, Z.; Zhang, H.T.; Wang, X. An antibacterial hemostatic AuNPs@corn stalk/chitin composite sponge with shape recovery for promoting wound healing. *Carbohydr. Polym.* **2022**, *296*, 119924. [CrossRef] [PubMed]
69. Zhan, X.; Yan, J.; Tang, H.; Xia, D.; Lin, H. Antibacterial Properties of Gold Nanoparticles in the Modification of Medical Implants: A Systematic Review. *Pharmaceutics* **2022**, *14*, 2654. [CrossRef]
70. More, P.R.; Zannella, C.; Folliero, V.; Foglia, F.; Troisi, R.; Vergara, A.; Franci, G.; De Filippis, A.; Galdiero, M. Antimicrobial Applications of Green Synthesized Bimetallic Nanoparticles from Ocimum basilicum. *Pharmaceutics* **2022**, *14*, 2457. [CrossRef]
71. Elgamily, H.M.; El-Sayed, H.S.; Abdelnabi, A. The Antibacterial Effect of Two Cavity Disinfectants against One of Cariogenic Pathogen: An In vitro Comparative Study. *Contemp. Clin. Dent.* **2018**, *9*, 457–462. [CrossRef] [PubMed]
72. Kuppusamy, P.; Ichwan, S.J.; Parine, N.R.; Yusoff, M.M.; Maniam, G.P.; Govindan, N. Intracellular biosynthesis of Au and Ag nanoparticles using ethanolic extract of *Brassica oleracea* L. and studies on their physicochemical and biological properties. *J. Environ. Sci.* **2015**, *29*, 151–157. [CrossRef]
73. Zhu, H.; Tan, J.; Qiu, J.; Wang, D.; Zhao, Z.; Lu, Z.; Huang, G.; Liu, X.; Mei, Y. Gold Nanoparticles Decorated Titanium Oxide Nanotubes with Enhanced Antibacterial Activity Driven by Photocatalytic Memory Effect. *Coatings* **2022**, *12*, 1351. [CrossRef]

74. Zheng, X.; Sun, J.; Li, W.; Dong, B.; Song, Y.; Xu, W.; Zhou, Y.; Wang, L. Engineering nanotubular titania with gold nanoparticles for antibiofilm enhancement and soft tissue healing promotion. *J. Electroanal. Chem.* **2020**, *871*, 114362. [CrossRef]
75. Takanche, J.S.; Kim, J.E.; Kim, J.S.; Lee, M.H.; Jeon, J.G.; Park, I.S.; Yi, H.K. Chitosan-gold nanoparticles mediated gene delivery of c-myb facilitates osseointegration of dental implants in ovariectomized rat. *Artif. Cells Nanomed. Biotechnol.* **2018**, *46*, S807–S817. [CrossRef] [PubMed]
76. Su, C.; Huang, K.; Li, H.-H.; Lu, Y.-G.; Zheng, D.-L. Antibacterial Properties of Functionalized Gold Nanoparticles and Their Application in Oral Biology. *J. Nanomater.* **2020**, *2020*, 5616379. [CrossRef]
77. Liu, X.; Tan, N.; Zhou, Y.; Wei, H.; Ren, S.; Yu, F.; Chen, H.; Jia, C.; Yang, G.; Song, Y. Delivery of antagomiR204-conjugated gold nanoparticles from PLGA sheets and its implication in promoting osseointegration of titanium implant in type 2 diabetes mellitus. *Int. J. Nanomed.* **2017**, *12*, 7089–7101. [CrossRef] [PubMed]
78. Moon, K.S.; Park, Y.B.; Bae, J.M.; Oh, S. Near-infrared laser-mediated drug release and antibacterial activity of gold nanorod–sputtered titania nanotubes. *J. Tissue Eng.* **2018**, *9*, 1–9. [CrossRef] [PubMed]
79. Elkhidir, Y.; Lai, R.; Feng, Z. The impact of photofunctionalized gold nanoparticles on osseointegration. *Heliyon* **2018**, *4*, e00662. [CrossRef]
80. Takanche, J.S.; Kim, J.E.; Jang, S.; Yi, H.K. Insulin growth factor binding protein-3 enhances dental implant osseointegration against methylglyoxal-induced bone deterioration in a rat model. *J. Periodontal Implant. Sci.* **2022**, *52*, 155–169. [CrossRef]
81. Tijana, A.; Valentina, V.; Nataša, T.; Miloš, H.M.; Suzana, G.A.; Milica, B.; Yoshiyuki, H.; Hironori, S.; Ivanič, A.; Rebeka, R. Mechanical properties of new denture base material modified with gold nanoparticles. *J. Prosthodont. Res.* **2021**, *65*, 155–161. [CrossRef] [PubMed]
82. Zhang, M.; Liu, X.; Xie, Y.; Zhang, Q.; Zhang, W.; Jiang, X.; Lin, J. Biological Safe Gold Nanoparticle-Modified Dental Aligner Prevents the Porphyromonas gingivalis Biofilm Formation. *ACS Omega* **2020**, *5*, 18685–18692. [CrossRef]
83. Golub, D.; Ivanič, A.; Majerič, P.; Tiyyagura, H.R.; Anžel, I.; Rudolf, R. Synthesis of Colloidal Au Nanoparticles through Ultrasonic Spray Pyrolysis and Their Use in the Preparation of Polyacrylate-AuNPs' Composites. *Materials* **2019**, *12*, 3775. [CrossRef]
84. Hashimoto, M.; Kawai, K.; Kawakami, H.; Imazato, S. Matrix metalloproteases inhibition and biocompatibility of gold and platinum nanoparticles. *J. Biomed. Mater. Res. A* **2016**, *104*, 209–217. [CrossRef] [PubMed]
85. Bukovinszky, K.; Szalóki, M.; Csarnovics, I.; Bonyár, A.; Petrik, P.; Kalas, B.; Daróczi, L.; Kéki, S.; Kökényesi, S.; Hegedűs, C. Optimization of Plasmonic Gold Nanoparticle Concentration in Green LED Light Active Dental Photopolymer. *Polymers* **2021**, *13*, 275. [CrossRef] [PubMed]
86. Łyczek, J.; Bończak, B.; Krzymińska, I.; Giżyński, K.; Paczesny, J. Gold–oxoborate nanocomposite-coated orthodontic brackets gain antibacterial properties while remaining safe for eukaryotic cells. *J. Biomed. Mater. Res. Part. B Appl. Biomater.* **2023**, *111*, 996–1004. [CrossRef] [PubMed]
87. Nam, K.Y. Characterization and antifungal activity of the modified PMMA denture base acrylic: Nanocomposites impregnated with gold, platinum, and silver nanoparticles. *Nanobiomaterials Dent. Appl. Nanobiomaterials* **2016**, *11*, 309–336.
88. Yang, T.; Qian, S.; Qiao, Y.; Liu, X. Cytocompatibility and antibacterial activity of titania nanotubes incorporated with gold nanoparticles. *Colloids Surf. B Biointerfaces* **2016**, *145*, 597–606. [CrossRef] [PubMed]
89. Pan, J.; Deng, J.; Yu, L.; Wang, Y.; Zhang, W.; Han, X.; Camargo, P.H.C.; Wang, J.; Liu, Y. Investigating the repair of alveolar bone defects by gelatin methacrylate hydrogels-encapsulated human periodontal ligament stem cells. *J. Mater. Sci. Mater. Med.* **2019**, *31*, 3. [CrossRef]
90. Khaohoen, A.; Sornsuwan, T.; Chaijareenont, P.; Poovarodom, P.; Rungsiyakull, C.; Rungsiyakull, P. Biomaterials and Clinical Application of Dental Implants in Relation to Bone Density-A Narrative Review. *J. Clin. Med.* **2023**, *12*, 6924. [CrossRef]
91. Bapat, R.A.; Joshi, C.P.; Bapat, P.; Chaubal, T.V.; Pandurangappa, R.; Jnanendrappa, N.; Gorain, B.; Khurana, S.; Kesharwani, P. The use of nanoparticles as biomaterials in dentistry. *Drug Discov. Today* **2019**, *24*, 85–98. [CrossRef]
92. Li, X.; Wang, Y.; Huang, D.; Jiang, Z.; Lei, J.; Xiao, Y.; He, Z.; Luo, M. Nanomaterials Modulating the Fate of Dental-Derived Mesenchymal Stem Cells Involved in Oral Tissue Reconstruction: A Systematic Review. *Int. J. Nanomed.* **2023**, *18*, 5377–5406. [CrossRef] [PubMed]
93. Zhang, Y.; Wang, P.; Wang, Y.; Li, J.; Qiao, D.; Chen, R.; Yang, W.; Yan, F. Gold Nanoparticles Promote the Bone Regeneration of Periodontal Ligament Stem Cell Sheets Through Activation of Autophagy. *Int. J. Nanomed.* **2021**, *16*, 61–73. [CrossRef] [PubMed]
94. Moon, K.S.; Choi, E.J.; Bae, J.M.; Park, Y.B.; Oh, S. Visible Light-Enhanced Antibacterial and Osteogenic Functionality of Au and Pt Nanoparticles Deposited on $TiO_2$ Nanotubes. *Materials* **2020**, *13*, 3721. [CrossRef]
95. Bapat, R.; Chaubal, T.; Dharmadhikari, S.; Abdulla, A.; Bapat, P.; Alexander, A.; Dubey, S.; Kesharwani, P. Recent advances of gold nanoparticles as biomaterial in dentistry. *Int. J. Pharm.* **2020**, *586*, 119596. [CrossRef]
96. Augustine, R.; Hasan, A. Emerging applications of biocompatible phytosynthesized metal/metal oxide nanoparticles in healthcare. *J. Drug Deliv. Sci. Technol.* **2020**, *56*, 101516. [CrossRef]
97. Zhang, Q.; Hou, D.; Wen, X.; Xin, M.; Li, Z.; Wu, L.; Pathak, J.L. Gold nanomaterials for oral cancer diagnosis and therapy: Advances, challenges, and prospects. *Mater. Today Bio.* **2022**, *15*, 100333. [CrossRef]
98. Rai, M.; Ingle, A.P.; Birla, S.; Yadav, A.; Santos, C.A. Strategic role of selected noble metal nanoparticles in medicine. *Crit. Rev. Microbiol.* **2016**, *42*, 696–719. [CrossRef] [PubMed]
99. Kamath, K.; Nasim, I.; Rajeshkumar, S. Evaluation of the re-mineralization capacity of a gold nanoparticle-based dental varnish: An in vitro study. *J. Conserv. Dent.* **2020**, *23*, 390–394. [CrossRef] [PubMed]

100. Butler, J.; Handy, R.D.; Upton, M.; Besinis, A. Review of Antimicrobial Nanocoatings in Medicine and Dentistry: Mechanisms of Action, Biocompatibility Performance, Safety, and Benefits Compared to Antibiotics. *ACS Nano* **2023**, *17*, 7064–7092. [CrossRef] [PubMed]
101. Adokoh, C.K.; Quan, S.; Hitt, M.; Darkwa, J.; Kumar, P.; Narain, R. Synthesis and evaluation of glycopolymeric decorated gold nanoparticles functionalized with gold-triphenyl phosphine as anti-cancer agents. *Biomacromolecules* **2014**, *15*, 3802–3810. [CrossRef]
102. Jiang, T.; Miao, S.; Shen, J.; Song, W.; Tan, S.; Ma, D. Enhanced effects of antagomiR-3074-3p-conjugated PEI-AuNPs on the odontogenic differentiation by targeting FKBP9. *J. Tissue Eng.* **2023**, *14*, 1–17. [CrossRef] [PubMed]
103. Yu, Q.; Li, J.; Zhang, Y.; Wang, Y.; Liu, L.; Li, M. Inhibition of gold nanoparticles (AuNPs) on pathogenic biofilm formation and invasion to host cells. *Sci. Rep.* **2016**, *6*, 26667. [CrossRef] [PubMed]
104. Han, S.; Locke, A.K.; Oaks, L.A.; Cheng, Y.L.; Coté, G.L. Nanoparticle-based assay for detection of S100P mRNA using surface-enhanced Raman spectroscopy. *J. Biomed. Opt.* **2019**, *24*, 055001. [CrossRef]
105. Li, C.; Xian, J.; Hong, J.; Cao, X.; Zhang, C.; Deng, Q.; Qin, Z.; Chen, M.; Zheng, X.; Li, M.; et al. Dual photothermal nanocomposites for drug-resistant infectious wound management. *Nanoscale* **2022**, *14*, 11284–11297. [CrossRef] [PubMed]
106. Ray, R.R.; Pattnaik, S. Technological advancements for the management of oral biofilm. *Biocatal. Agric. Biotechnol.* **2024**, *56*, 103017. [CrossRef]
107. Lavaee, F.; Motamedifar, M.; Rafiee, G. The effect of photodynamic therapy by gold nanoparticles on Streptococcus mutans and biofilm formation: An in vitro study. *Lasers Med. Sci.* **2022**, *37*, 1717–1725. [CrossRef] [PubMed]
108. Abbai, R.; Mathiyalagan, R.; Markus, J.; Kim, Y.J.; Wang, C.; Singh, P.; Ahn, S.; Farh, M.E.A.; Yang, D.C. Green synthesis of multifunctional silver and gold nanoparticles from the oriental herbal adaptogen: Siberian ginseng. *Int. J. Nanomed.* **2016**, *11*, 3131–3143. [CrossRef]
109. Shareena Dasari, T.P.; Zhang, Y.; Yu, H. Antibacterial Activity and Cytotoxicity of Gold (I) and (III) Ions and Gold Nanoparticles. *Biochem. Pharmacol.* **2015**, *4*, 199. [CrossRef]
110. Nemt-Allah, A.A.; Ibrahim, S.H.; El-Zoghby, A.F. Marginal Integrity of Composite Restoration with and without Surface Pretreatment by Gold and Silver Nanoparticles vs Chlorhexidine: A Randomized Controlled Trial. *J. Contemp. Dent. Pract.* **2021**, *22*, 1087–1097. [CrossRef]
111. Sun, Y.; Xu, W.; Jiang, C.; Zhou, T.; Wang, Q.; Lan, A. Gold nanoparticle decoration potentiate the antibacterial enhancement of $TiO_2$ nanotubes via sonodynamic therapy against peri-implant infections. *Front. Bioeng. Biotechnol.* **2022**, *10*, 1074083. [CrossRef] [PubMed]
112. Basova, T.V.; Vikulova, E.S.; Dorovskikh, S.I.; Hassan, A.; Morozova, N.B. The use of noble metal coatings and nanoparticles for the modification of medical implant materials. *Mater. Des.* **2021**, *204*, 109672. [CrossRef]
113. Zong, C.; Bronckaers, A.; Willems, G.; He, H.; de Llano-Pérula, M.C. Nanomaterials for Periodontal Tissue Regeneration: Progress, Challenges and Future Perspectives. *J. Funct. Biomater.* **2023**, *14*, 290. [CrossRef] [PubMed]

Journal of
*Functional Biomaterials*

MDPI

*Article*

# Plasma Rich in Growth Factors Compared to Xenogenic Bone Graft in Treatment of Periodontal Intra-Osseous Defects—A Prospective, Comparative Clinical Study

**Sourav Panda [1,*], Sital Panda [2], Abhaya Chandra Das [1], Natalia Lewkowicz [3], Barbara Lapinska [4], Margherita Tumedei [5,6], Funda Goker [5], Niccolò Cenzato [5] and Massimo Del Fabbro [5,6,*]**

1 Department of Periodontics, Institute of Dental Sciences, Siksha 'O' Anusandhan University, Bhubaneswar 751002, Odisha, India; abhayadas@soa.ac.in
2 Research Associate, Institute of Dental Sciences, Siksha 'O' Anusandhan University, Bhubaneswar 751002, Odisha, India; drsitalpanda@gmail.com
3 Department of Periodontology and Oral Diseases, Medical University of Lodz, 251 Pomorska St., 92-213 Lodz, Poland; natalia.lewkowicz@umed.lodz.pl
4 Department of General Dentistry, Medical University of Lodz, 251 Pomorska St., 92-213 Lodz, Poland; barbara.lapinska@umed.lodz.pl
5 Department of Biomedical, Surgical and Dental Sciences, Università degli Studi di Milano, 20122 Milan, Italy; margherita.tumedei@unimi.it (M.T.); funda.goker@unimi.it (F.G.); niccolo.cenzato@unimi.it (N.C.)
6 Fondazione IRCCS Ca' Granda Ospedale Maggiore Policlinico, 20122 Milan, Italy
* Correspondence: sauravpanda@soa.ac.in (S.P.); massimo.delfabbro@unimi.it (M.D.F.)

**Abstract:** Background: Periodontal intra-bony defects are challenging conditions in dental practice, often requiring regenerative approaches for successful treatment. This clinical study aimed to compare the effectiveness of plasma rich in growth factors (PRGF) versus xenogenic bone graft (BXG) in addressing intra-bony defects. Methods: Forty patients aged between 30 and 50 years presenting with generalized periodontitis were included. The study assessed various parameters, including relative attachment level (RAL); probing pocket depth (PPD); gingival marginal level (GML); intra-bony defect depth (IBDD) at baseline, 3, and 6 months; and level of pain, post-operative bleeding, and swelling, as patient-reported outcomes during the first seven days post operation. Results: The results revealed that both PRGF and BXG treatments led to significant reductions in IBDD over the 6-month study period. PRGF demonstrated significant advantages in GML enhancement and post-operative pain management during the initial post-treatment days. However, BXG showed a significantly greater reduction in IBDD compared to PRGF. Post-operative bleeding and swelling levels were comparable between the two treatments. Conclusions: These findings underscore the efficacy of both PRGF and BXG in periodontal regeneration, with treatment decisions guided by patient-specific factors and clinical goals.

**Keywords:** intra-osseous defects; periodontal regeneration; plasma rich in growth factors; xenograft

## 1. Introduction

Periodontal disease is a widespread oral health concern, affecting millions of individuals worldwide and posing a significant challenge to dental and periodontal care [1]. It is characterized by chronic inflammation of the periodontal tissues, leading to the degradation of the alveolar bone that supports the teeth and ultimately resulting in the formation of intra-bony defects [2–4]. These defects, representing localized pockets and bone loss, are pivotal in the progression of periodontal disease as they compromise the structural integrity of the periodontium and are associated with tooth mobility and tooth loss [5–7]. Addressing intra-bony defects is, therefore, imperative for the restoration of periodontal health and function.

Periodontal healing involves restoring the health of the tissues surrounding teeth after therapy, which can result in either repair or regeneration [8]. Repair is the re-establishment of tissue continuity without fully restoring the original architecture, often resulting in a long junctional epithelium or scar tissue. In contrast, regeneration seeks to completely rebuild lost structures, leading to new bone, cementum, and periodontal ligament formation, thus restoring both the structure and function of the periodontium [9].

Periodontal regeneration refers to the complete restoration of the lost or damaged periodontal structures, mimicking the essential wound-healing processes [10]. While several surgical techniques have been employed to create the optimal environment for periodontal regeneration, open flap debridement (OFD) or access flap surgery has shown promise in adjunct to a variety of biomaterials [11–14]. Nonetheless, these conventional strategies have limitations in achieving complete periodontal regeneration [15].

True regeneration can only occur through activating specific periodontal ligament-derived cells within the remaining periodontium, capable of differentiating into fibroblasts, cementoblasts, and osteoblasts [16,17]. The presence of a scaffold, cellular lineage and, most importantly, signaling molecules is essential for any tissue regeneration [17]. The cellular lineage can be obtained from the existing periodontal structures, while blood clots and bone grafts offer scaffold support [18]. Nonetheless, a crucial component often missing in these wound-healing events is the presence of signaling molecules.

Various therapeutic modalities have been employed to manage these challenging intra-bony defects, among which periodontal regeneration techniques have gained substantial recognition for their potential to promote tissue healing and regeneration. Specific methods within the realm of true regeneration have emerged as promising candidates for the augmentation of hard tissue regeneration within intra-bony defects, like the use of barrier membranes [19–21], bone grafts [22,23], stem cells [24,25], growth factors, and platelet concentrates.

Platelet concentrates derive from the patient's own whole blood and include different products like platelet-rich plasma (PRP), plasma rich in growth factors (PRGF), and platelet-rich fibrin (PRF). Such autologous products harness the regenerative potential of platelets, playing a pivotal role in the field of periodontal therapy [26] by facilitating the healing of intra-bony defects and other periodontal conditions. PRGF has drawn increasing attention in the last 25 years due to its regenerative potential, thanks to a concentrated and biologically active portion of the patient's blood plasma. PRGF is rich in growth factors and cytokines, which are instrumental in various cellular processes, including hemostasis, tissue healing, and regeneration [27,28]. The application of PRGF in periodontal therapy offers a minimally invasive and autologous approach, potentially enhancing the body's natural regenerative capabilities.

Bone grafts include autografts, allografts, xenografts, and alloplastic biomaterials, each with distinct advantages and considerations [29]. They serve as scaffolds, promoting new bone formation through osteoconduction and, in some cases, osteoinduction. Osteoinduction stimulates immature cells to become bone-forming cells, osteoconduction provides a scaffold for new bone growth, and osteogenesis is the direct formation of new bone by osteoblasts. Together, these processes facilitate effective bone healing and regeneration [29]. Xenogenic bone grafts consist of bone materials typically sourced from bovine or porcine origins [30,31]. The use of xenogenic bone grafts provides a clinically established and readily available alternative for enhancing periodontal tissue regeneration through osteoconduction.

Both PRGF and xenogenic bone grafts alone possess unique advantages in the context of regenerative periodontology, which raises the question of which intervention may provide superior clinical outcomes. Thus, the present clinical study aimed to address this fundamental question by directly comparing the efficacy of PRGF and xenogenic bone grafts in the treatment of periodontal intra-bony defects. Our research endeavors to elucidate the relative benefits, potential limitations, and clinical applicability of these two treatment modalities, contributing valuable insights to the optimization of regenerative approaches for patients suffering from periodontal disease. The null hypothesis of this study is that there is no significant difference in the effects of using PRGF and BXG on the periodontal parameters evaluated in this study.

## 2. Materials and Methods

The study was carried out at the out-patient department of Periodontics, Institute of Dental Sciences, Siksha 'O' Anusandhan University, India, between June 2020 and September 2022, as part of a PhD dissertation submitted to the University of Milan, Italy. The protocol was approved by the Institutional Review Board and the Ethics Council of Siksha 'O' Anusandhan University. The following selection criteria were employed for including patients:

Inclusion criteria:

1. Patients aged between 30 and 50 years;
2. Patients suffering from STAGE III periodontitis with grade A/B [32];
3. Patients presenting with 2- or 3-wall IBDs $\geq$ 3 mm deep measured from the alveolar crest to base of the defect;
4. Patients presenting with a probing depth (PD) $\geq$ 5 mm;
5. Patients who are systemically healthy and do not have any conditions that would contraindicate surgery.

Exclusion criteria:

1. Patients who underwent previous periodontal surgical treatment;
2. Patients presenting with interdental osseous craters;
3. Immuno-compromised patients;
4. Patients showing poor oral hygiene maintenance even after thorough scaling and root planing;
5. Women who are pregnant and lactating;
6. Patients on any antibiotic and/or steroid therapy within the last six months;
7. Patients presenting with teeth affected by peri-apical infection.

### 2.1. Study Design

This prospective, comparative clinical study was designed to compare PRGF with bovine xenogeneic graft (BXG) for surgical management of periodontal intra-osseous defects. Based on a pilot study carried out with similar groups on eight patients and the mean difference observed using relative attachment level (RAL) as the primary outcome; 16 per group, totaling 32 sites, was found to be the estimated sample size, with 80% as the statistical power and a significance level set at 5%.

### 2.2. Patient Characteristics

Fifty-two patients presenting with GRADE A/B and STAGE III periodontitis were initially enrolled for the present study. All patients underwent meticulous Phase I therapy. A re-evaluation was carried out after six weeks to ensure the patient's fitness to undergo surgery. Out of 52 enrolled patients, six failed to attend the re-evaluation appointment, and an additional six were excluded because they did not meet the predefined inclusion criteria. Finally, forty patients were randomized to receive one of the two treatments.

### 2.3. Outcomes

The clinical, radiological, and patient-reported outcomes were assessed at baseline and at various time points. The primary outcome of this study is RAL, and all other outcomes assessed were regarded as secondary. The details of each outcome and their methods of assessment are described below.

#### 2.3.1. Clinical Outcomes

- Plaque index (PI)—[33];
- Gingival index (GI)—[34];
- Probing depth (PD), in mm;
- Relative attachment level (RAL), in mm;
- Gingival marginal level (GML), in mm.

#### 2.3.2. Radiological Outcomes

Intra-bony defect depth (IBDD) in mm; measured on radiographs. The measurement was made from the deepest point of the intra-osseous defect to the imaginary line joining the adjacent cementoenamel junctions.

Intra-oral periapical radiographs (IOPAs) were made using the long cone paralleling technique. Position-indicating film holders were used to ensure stability. Customized bite blocks made from the putty index of patients were created and stored to ensure consistent positioning of the IOPAs at each post-operative recall interval.

For standardized exposure of the radiographs, the exposure time was set at 0.8 s, with a voltage of 70 kV and a current of 8 mA. The paralleling technique was used to prevent image overlap in the tooth's interproximal areas. All radiographs were digitized using an 800 dpi scanner (HP Scanjet 3c/I, Hewlett Packard, Palo Alto, CA, USA).

#### 2.3.3. Patient-Reported Outcome Measures (PROMs)

The pain level experienced in the first seven days post-operatively from Day 1 to Day 7 was recorded using a visual analog scale (VAS) ranging from 0 to 100, where 0 corresponds to no pain, and 100 corresponds to severe pain.

The bleeding and swelling at the treated sited were recorded from Day 1 to Day 7 by asking the patients to rate them from 0 to 5 (0—Never, 1—Rarely, 2—Occasionally, 3—Quite Often, 4—Very Often)

### *2.4. Study Follow-Up*

The follow-up of the study was carried out for 6 months. All the clinical and radiological outcomes were assessed at baseline, 3, and 6 months. And the PROMs were recorded post surgery until 1 week.

### *2.5. Randomization and Blinding*

Randomization was carried out using the coin toss method soon before starting the surgical phase. In this way, allocation concealment was ensured. The study was single-blinded, as the assessor (Si.P.) was masked at all time points. Neither the surgical operator nor the patient could be blinded due to the nature of the two treatments.

### *2.6. Surgical Procedure (Figure 1)*

All the periodontal surgery procedures were carried out by an experienced periodontist (S.P.). The patients were given a pre-procedural mouth rinse of 0.2% Chlorhexidine Gluconate (Hexidine, ICPA pharma, Mumbai, India). The surgical sites were anesthetized locally by administering 2% lignocaine (Lignox, Warren pharma, India).

Following adequate anesthesia, open access flap surgery was planned. A combination of crevicular and vertical incisions were placed to retract the flap and gain access to the defect. The defect was thoroughly degranulated using curettes (Standard Graceys. HuFriedy Group). The defect site was irrigated with saline and any remaining granulation tissues were completely removed. The defect site preparation was meticulously carried out to receive the active substitute.

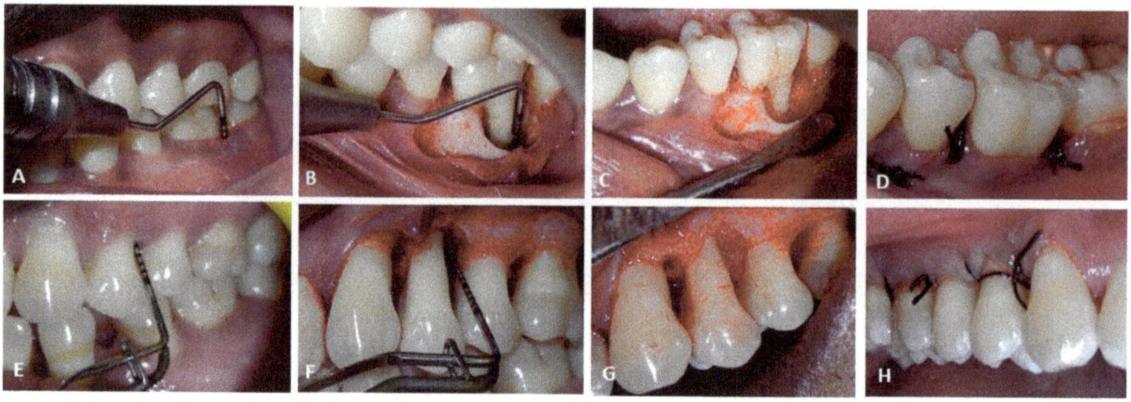


**Figure 1.** Surgical Procedure: (**A–D**) for BXG group; (**E–H**) for PRGF group.

### *2.7. For PRGF Group*

#### 2.7.1. Preparation of PRGF

First, 9 mL of the patient's blood was collected with 3.8% trisodium citrate as the anticoagulant. Once collected, the blood was subjected to centrifugation at 460× *g*. Centrifugation separates the blood into three distinct layers: the bottom fraction containing red blood cells, the middle fraction, also known as the "buffy coat", containing white blood cells, and the upper fraction, plasma, is rich in platelets and growth factors.

The upper fraction was divided into two fractions: F1 (Fraction 1) and F2 (Fraction 2). F2 was isolated and transferred to sterile tubes, and was then activated using 10% calcium chloride, to form the PRGF gel.

#### 2.7.2. Placement of PRGF into Defect Site

The obtained PRGF gel was then placed into the intra-bony defect and condensed until the whole defect was filled.

### *2.8. For BXG Group*

The infra-bony defects were packed and condensed up to an optimal level with the mixture of BXG (Bio-Oss, Geistlich Pharma, Switzerland) after proper debridement.

Single interrupted sutures using 6-0 monofilament suture (Ethicon) were placed to stabilize and secure the flap. The periodontal dressing was applied.

Patients were instructed with analgesics (paracetamol 500 mg) thrice a day for the next three days. Proper post-surgical instructions were provided to all patients, asking them to refrain from brushing at the operated area for the next seven days.

### *2.9. Statistical Analysis*

SPSS software version 26.0 (IBM, New York, NY, USA) was used to analyze the data. The Kolmogorov–Smirnov test and the Shapiro–Wilk test were used to check the normality of the quantitative data distributions. Based on the normality, Student's t-test was used to analyze the parametric data to compare outcomes between the groups, and the Mann–Whitney U test was used for the non-parametric set of data. A *p*-value of 0.05 was set as the level of significance.

## 3. Results

A total of 52 patients were assessed for eligibility, out of which 40 patients met the inclusion criteria and were randomly allocated to one of the two groups. No patients were lost to follow-up. A total of 20 patients in each group received the allocated intervention with either PRGF or BXG for management of intra-bony defects. (Figure 2). The demographic details of the patients are provided in Table 1.

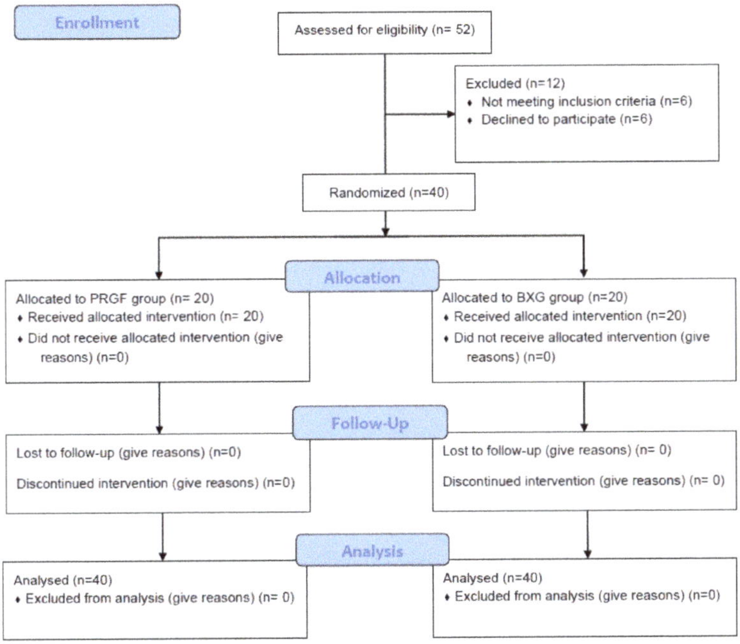


**Figure 2.** Flow diagram showing recruitment and follow-up of patients.

**Table 1.** Demographic data of both groups.

| | | Treatment Groups | | Total N | $p$-Value |
|---|---|---|---|---|---|
| | | PRGF | BXG | | |
| Sex | Female | 12 | 12 | 24 | |
| | Male | 8 | 8 | 16 | 1.000 [a] |
| Total | | 20 | 20 | 40 | |
| Age (in Years) | | $52 \pm 9$ | $53 \pm 10$ | 40 | 0.679 [b] |
| Smoker | N | 15 | 11 | 26 | 0.185 [a] |
| | Y | 5 | 9 | 14 | |
| Total | | 20 | 20 | 40 | |
| Walled_defect | 2 | 3 | 7 | 10 | 0.154 [a] |
| | 3 | 14 | 8 | 22 | |
| | Combined | 3 | 5 | 8 | |
| Total | | 20 | 20 | 40 | |

PRGF = plasma rich in growth factors; BXG = bovine xenogeneic graft; [a] Chi-square test; [b] Student's test.

## 3.1. Clinical Parameters

The PRGF group showed a significant improvement in GML at 3 and 6 months compared to the BXG group, indicating a potential advantage of PRGF in this aspect of periodontal health. However, other parameters such as PPD, RAL, PI, and GI did not show significant differences between the two groups (Table 2).

**Table 2.** Clinical parameters at all time points.

| Outcomes | Time Points | N | PRGF | BXG | *p*-Value |
|---|---|---|---|---|---|
| PPD (mm) | Baseline | 20 | 6.55 ± 1.05 | 6.55 ± 1.19 | 1.000 [a] |
| | 3 months | 20 | 4.45 ± 0.69 | 4.70 ± 0.86 | 0.398 [b] |
| | 6 months | 20 | 3.65 ± 0.81 | 3.90 ± 0.91 | 0.602 [b] |
| RAL (mm) | Baseline | 20 | 12.60 ± 1.50 | 12.10 ± 1.25 | 0.260 [a] |
| | 3 months | 20 | 10.95 ± 1.32 | 11.10 ± 1.45 | 0.718 [b] |
| | 6 months | 20 | 9.80 ± 1.36 | 9.95 ± 1.28 | 0.721 [a] |
| GML (mm) | Baseline | 20 | 1.90 ± 0.79 | 1.90 ± 0.72 | 0.977 [b] |
| | 3 months | 20 | 1.05 ± 0.69 | 1.60 ± 0.60 | 0.010 [b] |
| | 6 months | 20 | 0.85 ± 0.75 | 1.40 ± 0.88 | 0.041 [b] |
| PI | Baseline | 20 | 0.80 ± 0.24 | 0.93 ± 0.23 | 0.046 [b] |
| | 3 months | 20 | 0.62 ± 0.15 | 0.66 ± 0.15 | 0.414 [b] |
| | 6 months | 20 | 0.54 ± 0.15 | 0.51 ± 0.16 | 0.565 [a] |
| GI | Baseline | 20 | 1.75 ± 0.44 | 6.55 ± 1.19 | 0.035 [b] |
| | 3 months | 20 | 0.95 ± 0.22 | 0.80 ± 0.62 | 0.414 [b] |
| | 6 months | 20 | 0.75 ± 0.44 | 0.65 ± 0.49 | 0.602 [b] |

[a] Student's *t* test [b] Mann–Whitney U test.

The results indicate that both the PRGF and BXG treatment groups experienced reductions in PPD over the study period. In general, the differences between the study groups were not statistically significant. However, the PRGF group exhibited statistically significant improvements in RAL and GML at the 3-month follow-up, indicating a potential advantage of PRGF in accelerating the improvement in the RAL and GML (Table 3).

**Table 3.** Change at 3 and 6 months.

| Outcomes | Change | N | PRGF | BXG | *p*-Value |
|---|---|---|---|---|---|
| PPD (mm) | Baseline-3 months | 20 | −2.10 ± 0.91 | −1.85 ± 0.99 | 0.461 [a] |
| | Baseline-6 months | 20 | −2.90 ± 1.37 | −2.65 ± 0.99 | 0.738 [a] |
| RAL (mm) | Baseline-3 months | 20 | −1.65 ± 0.59 | −1.00 ± 0.79 | 0.007 [a] |
| | Baseline-6 months | 20 | −2.80 ± 0.77 | −2.15 ± 1.04 | 0.076 [a] |
| GML (mm) | Baseline-3 months | 20 | −0.85 ± 0.49 | −0.030 ± 0.57 | 0.009 [a] |
| | Baseline-6 months | 20 | −1.05 ± 0.39 | −0.50 ± 0.76 | 0.024 [a] |

[a] Wilcoxon signed rank test.

The reduction in the gingival recession was 2.8 times higher in the PRGF group compared to the BXG group at the end of 3 months and 2.1 times higher at the 6-month follow-up. Similarly, RAL gain was found to be 1.65 times greater in the PRGF group than the BXG group (Figure 3).

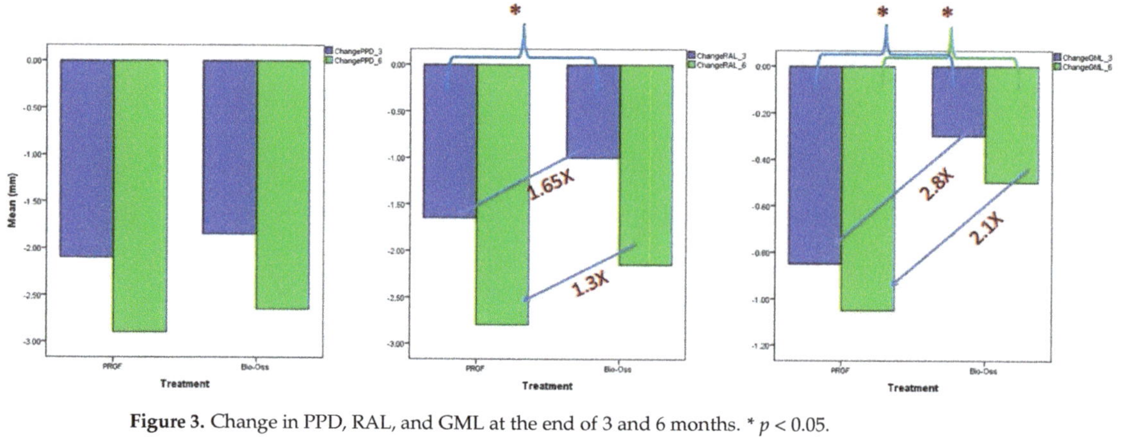


**Figure 3.** Change in PPD, RAL, and GML at the end of 3 and 6 months. * $p < 0.05$.

### 3.2. Radiological Parameters

The average reduction in the IBDD was 1.02 mm and 1.38 mm in the PRGF and BXG groups, respectively ($p = 0.043$) (Table 4).

**Table 4.** Intra-bony defect depth at baseline and 6 months in mm.

| Treatment | | IBDD-Baseline | IBDD 6 Months | Change at 6 Months |
|---|---|---|---|---|
| PRGF | N | 20 | 20 | 20 |
| | Mean | 4.39 | 3.37 | 1.02 |
| | SD | 0.97 | 0.88 | 0.43 |
| BXG | N | 20 | 20 | 20 |
| | Mean | 4.46 | 3.08 | 1.38 |
| | SD | 0.70 | 0.69 | 0.62 |
| *p*-value | | 0.183 [a] | 0.801 [a] | 0.261 [a] |

[a] Student's *t* test.

### 3.3. Patient-Reported Outcomes (PROMs)

Both the PRGF and BXG groups experienced a reduction in pain over the post-treatment period, with pain levels decreasing as the days progressed. While there were no statistically significant differences in pain levels between the two groups on the first and second days, significant differences emerged on the third, fifth, and seventh days, favoring the PRGF group (Table 5).

**Table 5.** Level of pain between both groups during week 1.

| Treatment | | Pain_D1 | Pain_D2 | Pain_D3 | Pain_D4 | Pain_D5 | Pain_D6 | Pain_D7 |
|---|---|---|---|---|---|---|---|---|
| PRGF (20) | Mean | 7.50 | 6.50 | 4.00 | 3.00 | 1.00 | 0.25 | 0.00 |
| | SD | 14.82 | 12.68 | 12.31 | 9.23 | 3.08 | 1.12 | 0.00 |
| BXG (20) | Mean | 22.00 | 20.00 | 16.50 | 12.50 | 8.00 | 7.00 | 5.25 |
| | SD | 29.31 | 25.55 | 29.25 | 26.33 | 18.24 | 15.59 | 10.94 |
| *p*-value | | 0.183 [a] | 0.117 [a] | 0.036 [a] | 0.112 [a] | 0.177 [a] | 0.064 [a] | 0.019 [a] |

[a] Mann–Whitney U test.

The differences in post-operative bleeding between the PRGF and BXG groups were not statistically significant over the first week, except for a slight significance on the first day. Post-operative swelling also did not significantly differ between the two groups on any of the days post treatment.

## 4. Discussion

Periodontal regeneration in the context of intra-bony defects represents a crucial aspect of periodontal therapy, aiming to restore the lost periodontal tissues and prevent further disease progression [10,35,36]. In this clinical study, we explored the efficacy of two prominent regenerative approaches—plasma rich in growth factors and xenogenic bone graft—in the treatment of periodontal intra-bony defects.

The initial comparison of demographic and clinical characteristics between the PRGF and BXG groups showed a similar distribution of sex, age, smoking status, and the features of walled defects. This suggests that the randomization process was effective in creating comparable treatment groups, reducing potential sources of bias.

Our primary objective was to evaluate the regenerative potential of PRGF compared to BXG in treating periodontal intra-bony defects. The results indicated that both treatment modalities led to improvements in clinical parameters such as probing depth reduction and clinical attachment level gain. These findings align with the existing literature [37,38] that recognizes the regenerative capabilities of both autologous platelet concentrates and bone substitutes in periodontal therapy.

PRGF is derived from autologous blood and offers promising sources of an array of growth factors. PRGF contains essential growth factors, including PDGF-AB, TGF β1, and VEGF, capable of stimulating cell proliferation, matrix remodeling, and angiogenesis [39]. Research has demonstrated the positive effects of PRGF on bone regeneration, including its application in rabbit calvarial bone defects [40], tibial bone defects [41], peri-implant bone healing [42,43], periodontal pockets [44], and human extraction socket healing [45–47]. PRGF has also shown promise in treating human periodontal defects, including grade 2 furcation defects [48,49], which significantly improved attachment levels and reduced defect depth and extent.

Xenografts are derived from non-human sources, most commonly bovine or porcine origin, and have gained prominence in the field of periodontal regeneration. These graft materials are processed to eliminate potential immunogenicity and pathogenicity concerns, rendering them biocompatible and safe for clinical use. Xenografts offer a versatile solution for periodontal therapy, particularly in cases where autologous grafts may be limited, such as insufficient donor site availability [50].

Xenografts facilitate periodontal regeneration through several mechanisms, including osteoconduction and providing a scaffold for new bone formation. The graft materials serve as a matrix for bone-forming cells to adhere to, proliferate, and produce new bone tissue [51]. Moreover, some xenografts possess osteo-inductive properties, further stimulating the differentiation of progenitor cells into osteoblasts, thereby promoting bone formation within intra-bony defects [52]. Therefore, BXGs have established themselves as a valuable resource in periodontal regeneration, offering an effective solution for the management of intra-bony defects.

However, it is important to note that in the present study there were no statistically significant differences between the two treatment groups in terms of probing depth reduction and relative attachment level gain. This suggests that, within the study's limitations and the specific patient population, PRGF and BXG have comparable efficacy in promoting periodontal regeneration in intra-bony defects. These results are in accordance with previous studies that have reported the effectiveness of both treatment modalities.

The observed increase in gingival marginal level for the PRGF group compared to the BXG group at the end of the follow-up period is an interesting finding with potential clinical significance. The presence of creeping attachment in the PRGF group suggests that the treatment has not only been effective in filling intra-bony defects but has also had a positive influence on the reattachment of periodontal tissues, such as the connective tissue and gingival margin. This result suggests that PRGF may have a more favorable impact on the gingival margin's position, which is an essential aspect of periodontal health and esthetics. The improvement in gingival marginal level in the PRGF group could be attributed to the regenerative potential of PRGF, which contains a concentrated and biologically active portion of the patient's blood plasma, rich in growth factors and cytokines [39]. These components are known to stimulate tissue healing and regeneration, including the regeneration of periodontal tissues. However, it is crucial to interpret this result within the context of the entire study. The statistical significance, clinical relevance, and sustainability of this increase in gingival marginal level should be thoroughly analyzed.

While there were no statistically significant differences in pain levels between the two groups on the first and second days, significant differences emerged on the third, fifth, and seventh days, favoring the PRGF group. This suggests that the PRGF group reported lower pain levels after the second day of surgery. It is important to consider that, while there are statistically significant differences on certain days, the clinical significance of these differences may vary, and the overall patient experience should be taken into account. Pain perception can be influenced by various factors, including individual pain thresholds and the specific surgical techniques used. Therefore, these findings should be treated with caution.

The radiographic method was chosen to assess intra-bony defect depth because of its non-invasive nature and its ability to provide a comprehensive view of the bone structure. Radiographs are widely used in periodontal studies for visualizing bone loss and defect morphology, which are essential for evaluating treatment outcomes. However, it is important to recognize that the accuracy of this method can be affected by various factors, including the resolution of the imaging system, the angle of the radiographic projection, and the calibration of measurements. To minimize these shortcomings, standardization of the radiographic technique is crucial. This includes consistent imaging protocols, positioning of the patient, and calibration of measurements, which can help ensure more accurate and reliable assessments.

Both the PRGF and BXG group were found to be effective in reducing IBDD over the 6-month study period. It was found that the change in IBDD was significantly greater in the BXG group compared to the PRGF group. However, PRGF represents a unique approach because it consists of autologous plasma rich in growth factors, which stimulates tissue healing and regeneration directly from the patient's own biological resources. This key distinction means that what we observe with PRGF is not merely a biomaterial filling, but rather a reflection of true tissue healing and regeneration. This insight underscores the importance of interpreting study results within the context of the treatment's biological mechanisms and the potential for radiopacity to influence measurements when using biomaterials like BXG.

The outcomes of this study have several clinical implications. First, clinicians can confidently consider both PRGF and BXG as viable options, depending on patient preferences, clinical circumstances, and available resources. The lack of statistically significant differences between these treatments suggests that the choice may be based on individual patient factors and clinical considerations.

In the context of our study comparing the effectiveness of PRGF and BXG in periodontal regeneration for intra-bony defects, several potential confounding factors demand consideration. These variables, if left unaddressed, have the potential to influence the study outcomes. These confounding factors include patient compliance with post-treatment care and follow-up, variations in the baseline severity of periodontal disease, discrepancies in periodontal diagnosis and disease progression rates, concomitant medication usage, diverse oral hygiene practices, the presence of smokers in each treatment group, previous dental procedures undergone, socioeconomic disparities among participants, operator skill and experience, and differences in systemic health. Controlling for these confounding factors via our robust study design, strict inclusion and exclusion criteria, randomization, and appropriate statistical analysis is imperative to ensure the reliability and validity of our study's conclusions regarding the relative efficacy of PRGF and BXG in achieving successful periodontal regeneration in intra-bony defects.

The comparable outcomes observed with PRGF and BXG may stem from their overlapping regenerative capabilities, as both aim to enhance bone healing, albeit through different mechanisms. PRGF promotes regeneration with growth factors that stimulate cell proliferation and osteoblast activity, while BXG acts as an osteoconductive scaffold for bone growth. In the two-wall and three-wall defects included in this study, both materials likely benefited from the natural regenerative potential of the pre-existing bony walls, which may have minimized differences. Also, variability in individual healing responses could have also played a role, further blurring potential differences. A longer follow-up and inclusion of diverse defect types might help clarify the long-term effectiveness of PRGF and BXG.

The limitations of this study may amount to a relatively small sample size and inclusion of a specific patient population, which may have influenced the generalizability of the results. Further research with larger and more diverse populations is warranted to strengthen the findings. Another limitation of this study is that the follow-up was limited to 6 months. While 6 months is generally sufficient for assessing periodontal regeneration and observing improvements in clinical parameters, longer follow-up periods could provide additional insights into the long-term stability and durability of the treatment outcomes over time, potentially revealing additional changes in bone density, clinical attachment, or probing depths that might not be evident in the shorter follow-up period.

## 5. Conclusions

In conclusion, both PRGF and BXG as substitutes were effective in promoting periodontal regeneration. Both treatment protocols proved equally effective with no statistical difference. However, PRGF exhibits a distinct advantage in enhancing gingival marginal levels in the short term and managing post-operative pain, as evidenced by early pain alleviation by the end of 3rd day post-operatively. These results suggest that clinicians can choose between PRGF and BXG based on patient-specific factors and clinical considerations, to tailor therapies to the individual needs of the patient. PRGF may be particularly advantageous for patients seeking faster soft tissue recovery. Future studies with larger sample size and more diverse population, and extended follow-up periods are warranted.

**Author Contributions:** Conceptualization, S.P. (Sourav Panda) and M.D.F.; methodology, S.P. (Sourav Panda) and M.T.; software, S.P. (Sital Panda); validation, S.P. (Sourav Panda), F.G., B.L. and M.D.F.; formal analysis, S.P. (Sital Panda), A.C.D.; investigation, S.P. (Sourav Panda) and N.C.; resources, S.P. (Sourav Panda), A.C.D.; data curation, S.P. (Sourav Panda), N.C. and M.T.; writing—original draft preparation, S.P. (Sourav Panda); writing—review and editing, B.L., F.G., M.D.F. and N.L.; visualization, S.P. (Sourav Panda); supervision, M.D.F.; project administration, N.L.; funding acquisition, B.L. All authors have read and agreed to the published version of the manuscript.

**Funding:** This study was partially funded by Italian Ministry of Health, Current research IRCCS.

**Institutional Review Board Statement:** The study was conducted in accordance with the Declaration of Helsinki and approved by the Scientific Review Board and Institutional Ethical Committee of Siksha 'O' Anusandhan University (SOAU/2020/IMS/I-11).

**Informed Consent Statement:** Informed consent was obtained from all subjects involved in the study.

**Data Availability Statement:** The original contributions presented in the study are included in the article, further inquiries can be directed to the corresponding authors.

**Conflicts of Interest:** The authors declare no conflicts of interest.

## References

1. Jain, N.; Dutt, U.; Radenkov, I.; Jain, S. WHO's Global Oral Health Status Report 2022: Actions, Discussion and Implementation. *Oral Diseases* **2024**, *30*, 73–79. [CrossRef] [PubMed]
2. Könönen, E.; Gursoy, M.; Gursoy, U.K. Periodontitis: A Multifaceted Disease of Tooth-Supporting Tissues. *J. Clin. Med.* **2019**, *8*, 1135. [CrossRef] [PubMed]
3. Gasner, N.S.; Schure, R.S. Periodontal Disease. In *StatPearls*; StatPearls Publishing: Treasure Island, FL, USA, 2023. [PubMed]
4. Muñoz-Carrillo, J.L.; Hernández-Reyes, V.E.; García-Huerta, O.E.; Chávez-Ruvalcaba, F.; Chávez-Ruvalcaba, M.I.; Chávez-Ruvalcaba, K.M.; Díaz-Alfaro, L.; Muñoz-Carrillo, J.L.; Hernández-Reyes, V.E.; García-Huerta, O.E.; et al. Pathogenesis of Periodontal Disease. In *Periodontal Disease—Diagnostic and Adjunctive Non-Surgical Considerations*; IntechOpen: London, UK, 2019; ISBN 978-1-78984-461-0. [CrossRef]
5. Petsos, H.; Ramich, T.; Nickles, K.; Dannewitz, B.; Pfeifer, L.; Zuhr, O.; Eickholz, P. Tooth Loss in Periodontally Compromised Patients: Retrospective Long-Term Results 10 Years after Active Periodontal Therapy—Tooth-Related Outcomes. *J. Periodontol.* **2021**, *92*, 1761–1775. [CrossRef] [PubMed]
6. Ghods, K.; Alaee, A.; Jafari, A.; Rahimi, A. Common Etiologies of Generalized Tooth Mobility: A Review of Literature. *J. Res. Dent. Maxillofac. Sci.* **2022**, *7*, 249–259. [CrossRef]
7. Bäumer, A.; Pretzl, B.; Cosgarea, R.; Kim, T.S.; Reitmeir, P.; Eickholz, P.; Dannewitz, B. Tooth loss in aggressive periodontitis after active periodontal therapy: Patient-related and tooth-related prognostic factors. *J. Clin. Periodontol.* **2011**, *38*, 644–651. [CrossRef]
8. Fraser, D.; Caton, J.; Benoit, D.S.W. Periodontal Wound Healing and Regeneration: Insights for Engineering New Therapeutic Approaches. *Front. Dent. Med.* **2022**, *3*, 815810. [CrossRef]
9. Position Paper: Periodontal Regeneration. *J. Periodontol.* **2005**, *76*, 1601–1622. [CrossRef]
10. Mancini, L.; Fratini, A.; Marchetti, E. Periodontal Regeneration. *Encyclopedia* **2021**, *1*, 87–98. [CrossRef]
11. Zhu, Y.; Tao, C.; Goh, C.; Shrestha, A. Innovative Biomaterials for the Treatment of Periodontal Disease. *Front. Dent. Med.* **2023**, *4*, 1163562. [CrossRef]
12. Varghese, J.; Rajagopal, A.; Shanmugasundaram, S. Role of Biomaterials Used for Periodontal Tissue Regeneration—A Concise Evidence-Based Review. *Polymers* **2022**, *14*, 3038. [CrossRef]

13. Alqahtani, A.M. Guided Tissue and Bone Regeneration Membranes: A Review of Biomaterials and Techniques for Periodontal Treatments. *Polymers* **2023**, *15*, 3355. [CrossRef] [PubMed]
14. Pietruska, M.; Skurska, A.; Pietruski, J.; Dolińska, E.; Arweiler, N.; Milewski, R.; Duraj, E.; Sculean, A. Clinical and Radiographic Evaluation of Intrabony Periodontal Defect Treatment by Open Flap Debridement Alone or in Combination with Nanocrystalline Hydroxyapatite Bone Substitute. *Ann. Anat. Anat. Anz.* **2012**, *194*, 533–537. [CrossRef] [PubMed]
15. Cirelli, J.A.; Fiorini, T.; Moreira, C.H.C.; de Molon, R.S.; Dutra, T.P.; Sallum, E.A. Periodontal Regeneration: Is It Still a Goal in Clinical Periodontology? *Braz. Oral Res.* **2021**, *35*, e09. [CrossRef] [PubMed]
16. Woo, H.N.; Cho, Y.J.; Tarafder, S.; Lee, C.H. The Recent Advances in Scaffolds for Integrated Periodontal Regeneration. *Bioact. Mater.* **2021**, *6*, 3328–3342. [CrossRef] [PubMed]
17. Ward, E. A Review of Tissue Engineering for Periodontal Tissue Regeneration. *J. Vet. Dent.* **2022**, *39*, 49–62. [CrossRef]
18. Pandit, N.; Malik, R.; Philips, D. Tissue Engineering: A New Vista in Periodontal Regeneration. *J. Indian Soc. Periodontol.* **2011**, *15*, 328–337. [CrossRef]
19. Vani, T.M.S.; Paramashivaiah, R.; Prabhuji, M.L.V.; Peeran, S.W.; Fageeh, H.; Tasleem, R.; Bahamdan, G.K.; Aldosari, L.I.N.; Bhavikatti, S.K.; Scardina, G.A. Regeneration of Intrabony Defects with Nano Hydroxyapatite Graft, Derived from Eggshell along with Periosteum as Barrier Membrane under Magnification—An Interventional Study. *Appl. Sci.* **2023**, *13*, 1693. [CrossRef]
20. Das, S.; Panda, S.; Nayak, R.; Mohanty, R.; Satpathy, A.; Das, A.C.; Kumar, M.; Lapinska, B. Predictability and Clinical Stability of Barrier Membranes in Treatment of Periodontal Intrabony Defects: A Systematic Review and Meta-Analysis. *Appl. Sci.* **2022**, *12*, 4835. [CrossRef]
21. Górski, B.; Jalowski, S.; Górska, R.; Zaremba, M. Treatment of Intrabony Defects with Modified Perforated Membranes in Aggressive Periodontitis: A 12-Month Randomized Controlled Trial. *Clin. Oral Investig.* **2018**, *22*, 2819–2828. [CrossRef]
22. Hanes, P.J. Bone replacement grafts for the treatment of periodontal intrabony defects. *Oral Maxillofac. Surg. Clin. N. Am.* **2007**, *19*, 499–512, vi. [CrossRef]
23. Mahajan, A.; Kedige, S. Periodontal Bone Regeneration in Intrabony Defects Using Osteoconductive Bone Graft versus Combination of Osteoconductive and Osteostimulative Bone Graft: A Comparative Study. *Dent. Res. J.* **2015**, *12*, 25–30. [CrossRef] [PubMed]
24. Ferrarotti, F.; Romano, F.; Gamba, M.N.; Quirico, A.; Giraudi, M.; Audagna, M.; Aimetti, M. Human Intrabony Defect Regeneration with Micrografts Containing Dental Pulp Stem Cells: A Randomized Controlled Clinical Trial. *J. Clin. Periodontol.* **2018**, *45*, 841–850. [CrossRef] [PubMed]
25. Chen, F.-M.; Gao, L.-N.; Tian, B.-M.; Zhang, X.-Y.; Zhang, Y.-J.; Dong, G.-Y.; Lu, H.; Chu, Q.; Xu, J.; Yu, Y.; et al. Treatment of Periodontal Intrabony Defects Using Autologous Periodontal Ligament Stem Cells: A Randomized Clinical Trial. *Stem Cell Res. Ther.* **2016**, *7*, 33. [CrossRef] [PubMed]
26. Dohan Ehrenfest, D.M.; Andia, I.; Zumstein, M.A.; Zhang, C.-Q.; Pinto, N.R.; Bielecki, T. Classification of Platelet Concentrates (Platelet-Rich Plasma-PRP, Platelet-Rich Fibrin-PRF) for Topical and Infiltrative Use in Orthopedic and Sports Medicine: Current Consensus, Clinical Implications and Perspectives. *Muscles Ligaments Tendons J.* **2014**, *4*, 3–9. [CrossRef]
27. Nishiyama, K.; Okudera, T.; Watanabe, T.; Isobe, K.; Suzuki, M.; Masuki, H.; Okudera, H.; Uematsu, K.; Nakata, K.; Kawase, T. Basic Characteristics of Plasma Rich in Growth Factors (PRGF): Blood Cell Components and Biological Effects. *Clin. Exp. Dent. Res.* **2016**, *2*, 96–103. [CrossRef]
28. Baca-Gonzalez, L.; Serrano Zamora, R.; Rancan, L.; González Fernández-Tresguerres, F.; Fernández-Tresguerres, I.; López-Pintor, R.M.; López-Quiles, J.; Leco, I.; Torres, J. Plasma Rich in Growth Factors (PRGF) and Leukocyte-Platelet Rich Fibrin (L-PRF): Comparative Release of Growth Factors and Biological Effect on Osteoblasts. *Int. J. Implant. Dent.* **2022**, *8*, 39. [CrossRef]
29. Zhao, R.; Yang, R.; Cooper, P.R.; Khurshid, Z.; Shavandi, A.; Ratnayake, J. Bone Grafts and Substitutes in Dentistry: A Review of Current Trends and Developments. *Molecules* **2021**, *26*, 3007. [CrossRef]
30. De Azambuja Carvalho, P.H.; de Oliveira Ciaramicolo, N.; Júnior, O.F.; Pereira-Filho, V.A. Clinical and Laboratorial Outcomes of Xenogeneic Biomaterials: Literature Review. *Front. Oral Maxillofac. Med.* **2023**, *5*, 8. [CrossRef]
31. Tahmasebi, E.; Alam, M.; Yazdanian, M.; Tebyanian, H.; Yazdanian, A.; Seifalian, A.; Mosaddad, S.A. Current Biocompatible Materials in Oral Regeneration: A Comprehensive Overview of Composite Materials. *J. Mater. Res. Technol.* **2020**, *9*, 11731–11755. [CrossRef]
32. Papapanou, P.N.; Sanz, M.; Buduneli, N.; Dietrich, T.; Feres, M.; Fine, D.H.; Flemmig, T.F.; Garcia, R.; Giannobile, W.V.; Graziani, F.; et al. Periodontitis: Consensus Report of Workgroup 2 of the 2017 World Workshop on the Classification of Periodontal and Peri-Implant Diseases and Conditions. *J. Periodontol.* **2018**, *89*, S173–S182. [CrossRef]
33. Silness, J.; Loe, H. Periodontal disease in pregnancy. II. correlation between oral hygiene and periodontal condtion. *Acta Odontol. Scand.* **1964**, *22*, 121–135. [CrossRef] [PubMed]
34. Loe, H.; Silness, J. Periodontal disease in pregnancy. I. prevalence and severity. *Acta Odontol. Scand.* **1963**, *21*, 533–551. [CrossRef] [PubMed]
35. Reynolds, M.A.; Kao, R.T.; Camargo, P.M.; Caton, J.G.; Clem, D.S.; Fiorellini, J.P.; Geisinger, M.L.; Mills, M.P.; Nares, S.; Nevins, M.L. Periodontal regeneration - intrabony defects: A consensus report from the AAP Regeneration Workshop. *J. Periodontol.* **2015**, *86*, S105–S107. [CrossRef]

36. Stavropoulos, A.; Bertl, K.; Spineli, L.M.; Sculean, A.; Cortellini, P.; Tonetti, M. Medium- and Long-term Clinical Benefits of Periodontal Regenerative/Reconstructive Procedures in Intrabony Defects: Systematic Review and Network Meta-analysis of Randomized Controlled Clinical Studies. *J. Clin. Periodontol.* **2021**, *48*, 410–430. [CrossRef]
37. Sadatmansouri, S.; Ayubian, N.; Pourseyediyan, T.; Saljughidarmian, S. Effect of Plasma Rich in Growth Factors on the Healing of Intrabony Defects in Human: A Comparative Clinical Trial. *J. Periodontol. Implant Dent.* **2010**, *2*, 5–11. [CrossRef]
38. Mathur, A.; Bains, V.K.; Gupta, V.; Jhingran, R.; Singh, G.P. Evaluation of Intrabony Defects Treated with Platelet-Rich Fibrin or Autogenous Bone Graft: A Comparative Analysis. *Eur. J. Dent.* **2015**, *9*, 100–108. [CrossRef]
39. Anitua, E.; Tejero, R.; Zalduendo, M.M.; Orive, G. Plasma Rich in Growth Factors Promotes Bone Tissue Regeneration by Stimulating Proliferation, Migration, and Autocrine Secretion in Primary Human Osteoblasts. *J. Periodontol.* **2013**, *84*, 1180–1190. [CrossRef]
40. Stumbras, A.; Kuliesius, P.; Darinskas, A.; Kubilius, R.; Zigmantaite, V.; Juodzbalys, G. Bone Regeneration in Rabbit Calvarial Defects Using PRGF and Adipose-Derived Stem Cells: Histomorphometrical Analysis. *Regen. Med.* **2020**, *15*, 1535–1549. [CrossRef]
41. Anitua, E.; Orive, G.; Pla, R.; Roman, P.; Serrano, V.; Andía, I. The Effects of PRGF on Bone Regeneration and on Titanium Implant Osseointegration in Goats: A Histologic and Histomorphometric Study. *J. Biomed. Mater. Res. Part A* **2009**, *91A*, 158–165. [CrossRef]
42. Del Fabbro, M.; Boggian, C.; Taschieri, S. Immediate Implant Placement Into Fresh Extraction Sites with Chronic Periapical Pathologic Features Combined with Plasma Rich in Growth Factors: Preliminary Results of Single-Cohort Study. *J. Oral Maxillofac. Surg.* **2009**, *67*, 2476–2484. [CrossRef]
43. Birang, R.; Tavakoli, M.; Shahabouei, M.; Torabi, A.; Dargahi, A.; Soolari, A. Investigation of Peri-Implant Bone Healing Using Autologous Plasma Rich in Growth Factors in the Canine Mandible After 12 Weeks: A Pilot Study. *Open Dent. J.* **2011**, *5*, 168–173. [CrossRef] [PubMed]
44. Panda, S.; Purkayastha, A.; Mohanty, R.; Nayak, R.; Satpathy, A.; Das, A.C.; Kumar, M.; Mohanty, G.; Panda, S.; Fabbro, M.D. Plasma Rich in Growth Factors (PRGF) in Non-Surgical Periodontal Therapy: A Randomized Clinical Trial. *Braz. Oral Res.* **2020**, *34*, e034. [CrossRef] [PubMed]
45. Anitua, E.; Murias-Freijo, A.; Alkhraisat, M.H.; Orive, G. Clinical, Radiographical, and Histological Outcomes of Plasma Rich in Growth Factors in Extraction Socket: A Randomized Controlled Clinical Trial. *Clin. Oral Investig.* **2015**, *19*, 589–600. [CrossRef] [PubMed]
46. Mozzati, M.; Gallesio, G.; di Romana, S.; Bergamasco, L.; Pol, R. Efficacy of Plasma-Rich Growth Factor in the Healing of Postextraction Sockets in Patients Affected by Insulin-Dependent Diabetes Mellitus. *J. Oral Maxillofac. Surg.* **2014**, *72*, 456–462. [CrossRef] [PubMed]
47. Farina, R.; Bressan, E.; Taut, A.; Cucchi, A.; Trombelli, L. Plasma Rich in Growth Factors in Human Extraction Sockets: A Radiographic and Histomorphometric Study on Early Bone Deposition. *Clin. Oral Implant. Res.* **2013**, *24*, 1360–1368. [CrossRef]
48. Jenabian, N.; Haghanifar, S.; Ehsani, H.; Zahedi, E.; Haghpanah, M. Guided Tissue Regeneration and Platelet Rich Growth Factor for the Treatment of Grade II Furcation Defects: A Randomized Double-Blinded Clinical Trial—A Pilot Study. *Dent. Res. J.* **2017**, *14*, 363–369. [CrossRef]
49. Lafzi, A.; Shirmohammadi, A.; Faramarzi, M.; Jabali, S.; Shayan, A. Clinical Comparison of Autogenous Bone Graft with and without Plasma Rich in Growth Factors in the Treatment of Grade II Furcation Involvement of Mandibular Molars. *J. Dent. Res. Dent. Clin. Dent. Prospect.* **2013**, *7*, 22–29. [CrossRef]
50. Chen, F.-M.; Jin, Y. Periodontal Tissue Engineering and Regeneration: Current Approaches and Expanding Opportunities. *Tissue Eng. Part B Rev.* **2010**, *16*, 219–255. [CrossRef]
51. Smith, B.T.; Shum, J.; Wong, M.; Mikos, A.G.; Young, S. Bone Tissue Engineering Challenges in Oral & Maxillofacial Surgery. *Adv. Exp. Med. Biol.* **2015**, *881*, 57–78. [CrossRef]
52. Battafarano, G.; Rossi, M.; De Martino, V.; Marampon, F.; Borro, L.; Secinaro, A.; Del Fattore, A. Strategies for Bone Regeneration: From Graft to Tissue Engineering. *Int. J. Mol. Sci.* **2021**, *22*, 1128. [CrossRef]

Journal of
*Functional Biomaterials*

MDPI

*Systematic Review*

# Application of Advanced Platelet-Rich Fibrin in Oral and Maxillo-Facial Surgery: A Systematic Review

**Marek Chmielewski [1], Andrea Pilloni [2] and Paulina Adamska [3,*]**

[1] Private Dental Practice, 14 Kolberga Street, 81-881 Sopot, Poland; machmielewski@proton.me
[2] Section of Periodontics, Department of Oral and Maxillo-Facial Sciences, Sapienza University of Rome, 00-185 Rome, Italy; andrea.pilloni@uniroma1.it
[3] Division of Oral Surgery, Medical University of Gdańsk, 7 Dębinki Street, 80-211 Gdańsk, Poland
* Correspondence: paulina.adamska@gumed.edu.pl

**Abstract: Background:** Advanced platelet-rich fibrin (A-PRF) is produced by centrifuging the patient's blood in vacuum tubes for 14 min at 1500 rpm. The most important component of A-PRF is the platelets, which release growth factors from their $\alpha$-granules during the clotting process. This process is believed to be the main source of growth factors. The aim of this paper was to systematically review the literature and to summarize the role of A-PRF in oral and maxillo-facial surgery. **Materials and Methods:** A systematic review was carried out, following the Preferred Reporting Items for Systematic Reviews and Meta-Analyses (PRISMA) guidelines (PROSPERO: CRD42024584161). **Results:** Thirty-eight articles published before 11 November 2024 were included in the systematic review. The largest study group consisted of 102 patients, and the smallest study group consisted of 10 patients. A-PRF was most often analyzed compared to leukocyte-PRF (L-PRF) or blood cloth. A-PRF was correlated with lower postoperative pain. Also, A-PRF was highlighted to have a positive effect on grafting material integration. A-PRF protected areas after free gingival graft very well, promoted more efficient epithelialization of donor sites and enhanced wound healing. **Conclusions:** Due to its biological properties, A-PRF could be considered a reliable addition to the surgical protocols, both alone and as an additive to bio-materials, with the advantages of healing improvement, pain relief, soft tissue management and bone preservation, as well as graft integration. However, to determine the long-term clinical implications and recommendations for clinical practice, more well-designed randomized clinical trials are needed in each application, especially those with larger patient cohorts, as well as additional blinding of personnel and long follow-up periods.

**Keywords:** advanced platelet-rich fibrin; A-PRF; autografts; dentistry; growth factors; wound healing

## 1. Introduction

Since the first clinical introduction of platelet-rich fibrin (PRF) in dentistry by Choukroun in 2001 [1–4], PRF has grown to be one of the most influential natural resources in regenerative dentistry. In 2014, advanced platelet-rich fibrin (A-PRF) was introduced and described by Ghanaati et al. [5] and Choukroun [6] as one of the most promising iterations.

So far, three generations of blood-derived platelet-rich preparations rich in growth factors have been identified:

I. platelet-rich plasma (PRP), plasma-rich in growth factors (PRGF);
II. platelet-rich fibrin;
III. advanced platelet-rich fibrin, advanced platelet-rich fibrin plus (A-PRF+), injectable platelet-rich fibrin (I-PRF), concentrated growth factor (CGF), titanium platelet-rich fibrin (T-PRF) and autologous fibrin glue (AFG) [7,8].

A-PRF (Figure 1) is derived from the patient's venous blood drawn prior to surgery (most commonly from the brachial vein) without the use of anticoagulants. The autogenous origin ensures no undesirable antigen reactions after graft placement and during integration. The gelatinous consistency and state are similar to PRF due to a similar process of obtaining it [9,10]. It consists of a fibrinous matrix with mixed platelets, leukocytes, macrophages, neutrophils, proteins, cytokines and growth factors present in the blood, which are densely trapped. The matrix ensures mechanical strength and serves as a binding agent for cytokines and growth factors. A looser structure provides a more even distribution of cytokines than PRF, and more interfibrous space allows for more cells present in the cloth [5,9,10].

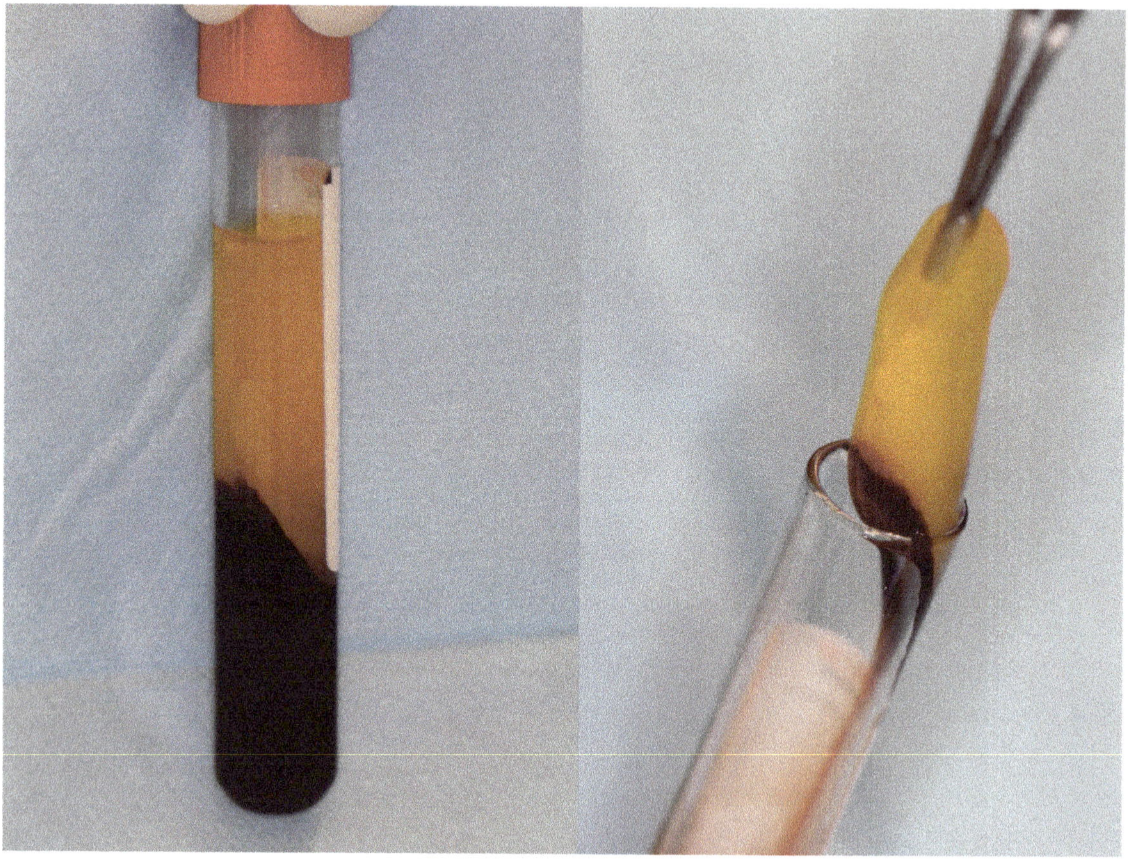

**Figure 1.** A-PRF clot in glass-coated plastic tubes.

Cytokines and growth factors play an important role in biochemical properties. A-PRF affects both soft and hard tissues, mainly due to the effects on tissue fibroblast regeneration [10,11] and osteoblasts [11]. The most important part of A-PRF are platelets that release growth factors from their α-granularities during clotting. This process is believed to be the main source of growth factors, such as platelet-derived growth factor (PDGF), vascular endothelial growth factor (VEGF), fibroblast growth factor (FGF), transforming growth factors (TGF-α and β), bone morphogenetic proteins (BMPs, such as BMP-2) and matrix metalloproteinases (MMPs, such as MMP-9). A-PRF has been shown to release more of these growth factors and MMPs than PRF due to its modified centrifugal processing. Research works indicate that A-PRF is currently the most favorable PRF-like material available, providing cytokines for up to 10 days [6,9,12–14]. Also, the addition of $Ca^{2+}$ can alter the amount of growth factors secreted into surrounding tissues [15].

In order to obtain A-PRF, the patient's blood has to be placed in a centrifuge and processed at 1500 rpm for 14 min, as opposed to the classical PRF at 3000 rpm for 12 min. There is also no anticoagulant, which gives A-PRF better biological properties compared to classical PRF. The lower centrifugation parameters of A-PRF compared to PRF or L-PRF do not allow the platelets to be pushed down the tube. The advantage of lower centrifuge speed in A-PRF preparation is the improvement of properties. Neutrophils can migrate to the fibrin matrix [6]. It is proven that in A-PRF in the peripherals of the cloth, there are platelets present. The difference in the processing can be responsible for more optimized, longer lasting and more even distribution and release of growth factors from A-PRF to the surrounding tissues, affecting tissue regeneration and maturation in comparison to PRF and L-PRF. The distribution of lymphocytes, macrophages and stem cells is greater in the proximal part of the cloth, whereas neutrophils are located mainly in the distal part [2,5,6]. It is postulated to reduce the formation time of A-PRF to 8 min, which further improves its biological properties. Further modifications of centrifugation led to the creation of I-PRF (700 rpm, 3 min), which has even more concentrated factors than advanced and leukocyte PRF. In comparison to PRF, it must be used within 20 min of preparation vs. 4 h for A-PRF. It is very important to maintain the speed of rotations per minute, time, relative centrifugal force, diameter and angulation of the rotor in the centrifuge, size and type of test tubes. Any change may lead to the incorrect production of blood-derived platelet-rich preparations rich in growth factors and loss of the related properties. Horizontal vs. fixed angle centrifugation is also important because horizontal centrifugation leads to four times higher cell concentration, which is evenly distributed across the top of the tube and is not damaged as much. The more hydrophilic the tube surface, the better the clot quality [2,5,6,16,17].

A-PRF is primarily used in surgical procedures. However, it can also be used in general oral surgery (filling the post-extraction socket, in treatment of alveolar osteitis, used to control bleeding–hemostatic effect), endodontics (in regenerative endodontic treatment (RET) of traumatized immature non-vital teeth), implantology (bone regeneration, socket preservation, alveolar ridge preservation, maxillary sinus augmentation), periodontics (in treatment recessions) and to enhance general wound healing (reduced pain, swelling or trismus) [2,18–40]. This makes A-PRF just as versatile as PRF. The aim of this paper was to collect and review the information about A-PRF, its role and its advantages in oral and maxillo-facial surgery.

## 2. Materials and Methods

### 2.1. Focused Question

The following focused question was defined: 'Does A-PRF provide better clinical outcomes than other materials used in exact oral and maxillo-facial procedures?' The PICO (population, intervention, comparison, outcome) was used:

P—at least 10 people qualified for the use of A-PRF;

I—dental procedures (i.e., tissue repair, socket management, sinus lifts) combined with the use of A-PRF as sole/combined bio-material;

C—defined approaches using A-PRF only or with other PRF types and other conventional methods;

O—soft and/or hard tissue reconstruction of the periodontium, alveolar bone or tooth structure.

### 2.2. Search Strategy

The search was conducted using PubMed, Scopus and Google Scholar web databases. The PRISMA guidelines (Preferred Reporting Items for Systematic Reviews and Meta-Analyses) were used [41]. The study protocol was registered with the PROSPERO (Prospective Register of Systematic Reviews, CRD42024584161). The search terms used were 'advanced platelet-rich fibrin', 'A-PRF', 'dentistry', 'oral surgery' and 'maxillo-facial surgery'. The last manuscript search was conducted on 11 November 2024.

### 2.3. Selection Criteria

#### 2.3.1. Inclusion Criteria

For the use of A-PRF in dentistry, only randomized clinical trials were used, with a minimum of 10 patients treated. Studies were selected from the databases according to the following criteria: (1) only human studies, (2) studies regarding A-PRF use in oral and maxillo-facial procedures, (3) studies carried out and published from the beginning of 2014 up to 11 November 2024, (4) studies published in English, (5) randomized clinical studies on a group of at least 10 people.

#### 2.3.2. Exclusion Criteria

The exclusion criteria were (1) studies conducted on animals or in vitro, (2) studies not available in full text, (3) group of patients less than 10, (4) studies written in language other than English.

### 2.4. Study Selection and Data Extraction

The retrieved publications were initially scanned in accordance with the selection criteria. Only publications fulfilling the inclusion criteria were taken into account. Duplicates from the databases were discarded. Upon screening the abstracts, selected articles were obtained in full text. If the abstract screening and title did not provide enough decisive information to include the article, further screening of the whole publication was carried out. Lastly, the full-text manuscripts were reviewed according to the selection criteria. Subsequently, the first author analyzed the publications and critically assessed the articles. In cases of uncertainty, the analysis was completed by a third author.

The data extracted included general publication characteristics (authors, year of publication, journal), case type, number of patients, outcome and complications. The analyzed data from the publications were divided according to the procedure and presented in tables for comparison.

### 2.5. Quality Assessment

In this article, the risk of bias was assessed using the Revised Cochrane risk of bias tool for randomized trials (RoB 2) [42,43]. These procedures were performed by the third and first authors. In case of disagreement, a consensus reading was made.

### 2.6. Statistical Analysis

Statistical analysis was performed using STATISTICA 13.3 (TIBCO, Palo Alto, CA, USA) licensed to the Medical University of Gdańsk. The number of studies, included publications and patients studied were quantitatively summarized.

A qualitative synthesis was conducted using the established criteria, summarizing the available research and analyzing the structure of PICO; the advantages, disadvantages, future research directions and the relationship with previous scientific reports were discussed. Quantitative analysis (meta-analysis) was not performed due to the heterogeneity of studies.

## 3. Results

### 3.1. Search Outcomes

After eliminating duplicates, 375 articles remained to be reviewed. After exclusion of duplicates, 261 articles remained. The screening of titles and abstracts excluded 140 articles. Ultimately, 38 articles were selected for systematic review. They were further divided into their respective category depending on the procedure carried out in the study (Figure 2). The first identified studies are from 2015. All randomized clinical studies are depicted in Table 1. The total number of patients analyzed in all studies was 1307 [2,19–40,44–60].

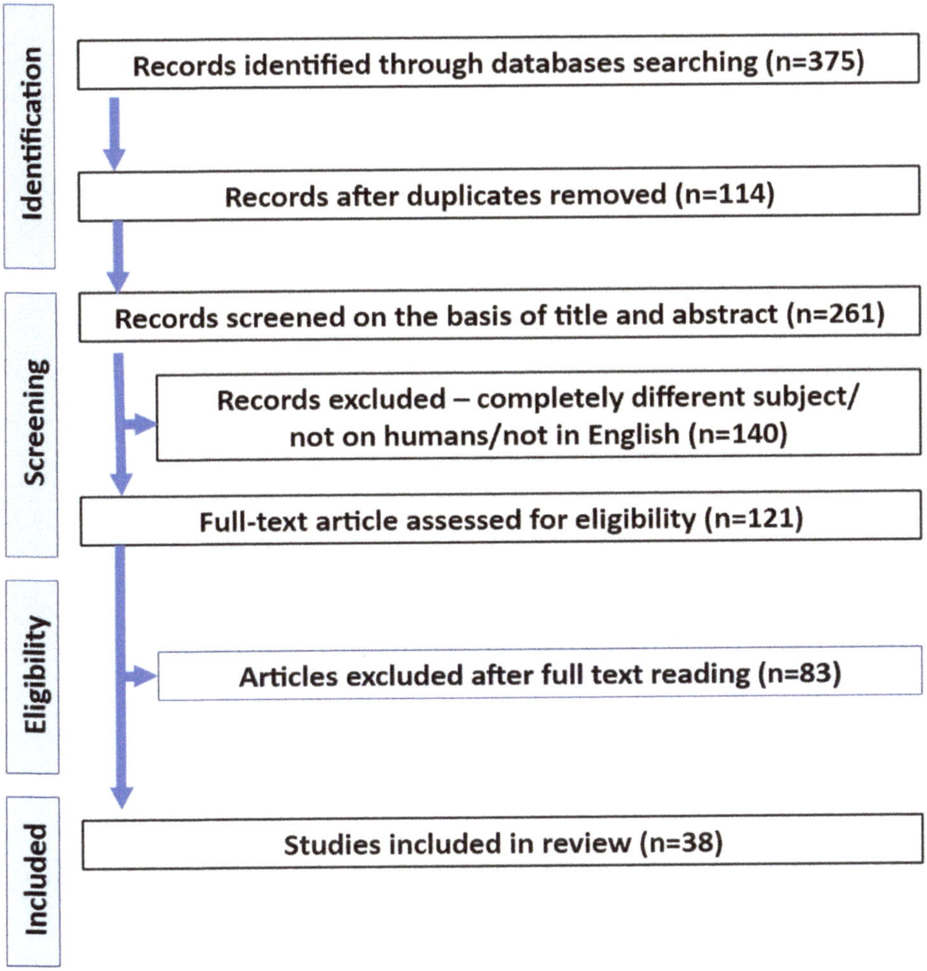


**Figure 2.** PRISMA workflow.

### 3.2. *Results of Individual Studies*

#### 3.2.1. A-PRF in Alveolar Ridge Preservation After Tooth Extraction

A-PRF is a material densely packed with growth factors and cytokines. This makes it a promising material for ridge preservation after tooth extraction. To date, there have been four randomized clinical trials (RCTs) that used A-PRF, attempting to hinder bone resorption after tooth extraction [30,32,39,48].

**Table 1.** Studies included in the qualitative analysis and extracted data.

| No | References | Aim of the Study | Number of Patients | Follow-Up | Results | Complications |
|---|---|---|---|---|---|---|
| 1 | Al-Barakani et al., 2024 [24] | The clinical effects of A-PRF or resorbable collagen membrane applied in the treatment of type I and II Miller class gingival recessions using the pin hole surgical technique. | 18 | 1, 2, 3, 4, 3 months | The treatment of recession using A-PRF in the pin hole surgical technique proved more effective than the application of resorbable collagen membrane in the pin hole method. In the case of the treatment of recession with the pin hole method and A-PRF simultaneously, a reduction in the level of postoperative pain was observed. | Not specified |
| 2 | Alhaj et al., 2018 [25] | Filling the resultant gap after immediate insertion of a mandibular molar implant with A-PRF + autograft mixture or autograft alone and comparing the outcomes. | 20 | Days 2 and 7; 3 and 6 months | After 6 months, the crystal bone decrease was more pronounced in the control bone, with statistical significance. A-PRF addition decreased swelling. A-PRF promoted faster regeneration. | Not specified |
| 3 | Alsahli et al., 2021 [26] | Comparing palatal free gingival graft and A-PRF as a material for patching uncovered implant sites during healing abutment placement and improving keratinization. | 15 | 1, 4, 8 weeks and 6 months | After 2 weeks of healing, the A-PRF group showed statistically significant improvement in keratinized tissue thickness, but the effect decreased over time. A-PRF improved the width of keratinized tissue, but with no significant advantage over free gingival graft. A-PRF was shown to decrease postoperative morbidity in comparison with free gingival graft. | One patient dropped out |
| 4 | Angelo et al., 2015 [27] | Maxillary implant placement after piezotome-enhanced subperiosteal tunnel technique with the use of bio-material with/without A-PRF membrane. | 82 | 6–7 months | The study suggests that A-PRF enhances biomechanic bone quality and allows for better and more consistent results with lower patient morbidity compared to traditional methods. | None |
| 5 | Bahammam, 2018 [28] | Patching free gingival graft sites with A-PRF and evaluating its impact on the donor site. | 24 | 1, 2, 3, 4 and 8 weeks | A-PRF is an effective bandage for free gingival graft donor site and helps in the early healing stages of soft tissues by promoting epithelialization. | Not specified |

**Table 1.** *Cont.*

| No | References | Aim of the Study | Number of Patients | Follow-Up | Results | Complications |
|---|---|---|---|---|---|---|
| 6 | Barakat et al., 2024 [29] | The clinical effects of A-PRF or connective tissue graft applied in the treatment of gingival black triangle using Han and Takei's method. | 32 | 1, 3, 6, 9 and 12 months | A-PRF and connective tissue graft had the same results in the interdental papilla treatment using Han and Takei's method. | Not specified |
| 7 | Brahma Prasad Chary et al., 2021 [20] | The evaluation of treatment outcome for early implant placement in sockets preserved using A-PRF at 6 and 8 weeks following atraumatic extraction. | 20 | 6–8 weeks | Better effects were achieved after 8 weeks (higher insertion torque values and predictable bone). | Not specified |
| 8 | Brancaccio et al., 2020 [2] | Extraction of four non-adjacent teeth with treatment using four different hemostatic procedures (sutures only, A-PRF+, HEM, L-PRF). | 102 | 2 weeks | HEM, A-PRF and leukocyte-PRF showed advantage over suturing alone. A-PRF presented a statistically significant advantage over HEM and fared best in bleeding reduction. Only L-PRF reduced the risk of incomplete healing compared to suturing alone with statistical significance. Hypertension and diabetes increased the risk of bleeding, and smoking and diabetes promoted delayed healing. | Not specified |
| 9 | Castro et al., 2021 [30] | Patching teeth sockets after multiple extractions in the front maxilla region with A-PRF or L-PRF and measuring the alveolar ridge dimension changes. | 21 | 3 months | Both PRF types could not counteract the progressing bone resorption after 3 months and yielded similar results. Both PRFs turned out to be superior in comparison with unassisted teeth sockets. | Not specified |
| 10 | Caymaz et al., 2018 [31] | Managing the socket after third molar extraction with the use of A-PRF and L-PRF. | 27 | - | A-PRF significantly lowered analgesic usage and the Visual Analog Scale compared to L-PRF, mainly in the first three days following surgery. There was no significant difference in terms of swelling and trismus. | Not specified |

**Table 1.** *Cont.*

| No | References | Aim of the Study | Number of Patients | Follow-Up | Results | Complications |
|---|---|---|---|---|---|---|
| 11 | Clark et al., 2018 [32] | Non-traumatic extraction with the use of A-PRF, A-PRF + FDBA, FDBA or blood clot for ridge preservation and further histomorphologic evaluation of the bone formed. | 40 | 3.75 months | The best results were achieved using A-PRF + FDBA. No significant difference between A-PRF and FDBA in terms of ridge dimension preservation was found. A-PRF and A-PRF + FDBA fared significantly better than blood cloth alone. Using A-PRF or A-PRF + FDBA resulted in formation of a denser trabecular structure. A-PRF also demonstrated the highest percentage of vital bone formation. | Not specified |
| 12 | Csifó-Nagy et al., 2021 [21] | Treating periodontal intrabony defects with A-PRF or EMD. | 18 | 6 months | In both groups, the FMBS decreased, and FMPS remained the same. In terms of pocket depth, gingival recession, clinical attachment level and bone sounding changes, A-PRF fared similarly to EMD, showing improvements compared to the baseline; thus, it can be concluded that A-PRF behaves as effectively as EMD in the surgical treatment of intrabony periodontal defects. | None |
| 13 | Dayashankara Rao et al., 2020 [33] | Performing secondary alveolar bone grafting using iliac bone graft alone or with a mixture of I- and A-PRF. | 30 | 3 and 6 months | The combination of I + A-PRF with iliac bone graft generated better results than using iliac bone graft alone, with good bone volume and lower chances of resorption. The periodontal status, mobility score and pocket depth improved in both groups, with no statistical significance. | The study group had 6.7% graft failure. The control group had 40% graft failure |
| 14 | Dragonas et al., 2023 [34] | Comparison of A-PRF and plasma rich in growth factors combined with DBBM during sinus lift augmentation. | 15 | 6 months | The mean percentage of mineralized bone after the healing period was higher in the group with growth factors, but there was no statistical significance in samples without growth factors. Adding A-PRF and PRGF to DBBM did not improve new bone formation in sinus lift. Neither platelet-rich preparation was better than the other in any of the parameters studied. | Not specified |

**Table 1.** *Cont.*

| No | References | Aim of the Study | Number of Patients | Follow-Up | Results | Complications |
|---|---|---|---|---|---|---|
| 15 | Ghonima et al., 2020 [35] | Regeneration of periodontal intrabony defects using BCP only or together with A-PRF. | 22 | 3, 6, 9 months | At 9 months, both groups had a significantly lower plaque index. At 3-, 6- and 9-month baselines, both groups significantly decreased the PD and gained CAL. No statistically significant difference was observed between A-PRF/BCP and BCP/saline groups, although the A-PRF group noted better results in PD reduction and CAL gain. | Not specified |
| 16 | Giudice et al., 2019 [36] | Extraction of four non-adjacent teeth with treatment using four different hemostatic procedures (sutures only, A-PRF+, HEM, L-PRF). | 40 | 1 and 2 weeks | A-PRF+ showed statistically significant bleeding reduction 30 min after extraction. In terms of patient preference and wound healing index, all types of plugs were similarly matched, although the L-PRF and A-PRF groups had higher percentage of complete closures compared to suturing and HEM after a 2-week period. | Not specified |
| 17 | Hartlev et al., 2020 [37] | Autogenous bone augmentation in future implant sites with additional use of A-PRF or DBBM and collagen membrane and analysis of vital bone formation. | 27 | 6 months | There were no significant differences between the control and test groups regarding vital bone and non-vital bone formation, the amount of blood vessels and soft tissues. | Two biopsies were discarded due to poor quality control group |
| 18 | Hartlev et al., 2021 [38] | Mandibular ramus block harvesting and lateral ridge augmentation with coverage of both sites with either A-PRF/resorbable collagen membrane or deproteinized bovine bone/resorbable collagen membrane. | 27 | 1 and 2 weeks | Both groups experienced low postoperative pain. The A-PRF group experienced lower pain perception, although statistically significant difference was only identified on the first postoperative day. | Changed sensation extra orally in the chin region, bone graft dehiscence in the recipient site control group, sensory disturbances at the recipient site |
| 19 | Ivanova et al., 2019 [39] | Extraction using A-PRF only or with FDBA and analyzing its effect on vital bone formation and ridge preservation. | 60 | 4 months | There were no significant differences between the use of allograft and A-PRF in terms of vertical bone resorption and vital bone creation. The use of both A-PRF and allograft outperformed the control group. | Not specified |

**Table 1.** *Cont.*

| No | References | Aim of the Study | Number of Patients | Follow-Up | Results | Complications |
|---|---|---|---|---|---|---|
| 20 | Jayadevan et al., 2021 [40] | The evaluation of A-PRF and PRF as a scaffold in the regenerative endodontic treatment of traumatized immature non-vital permanent anterior teeth. | 28 | 13 months | A-PRF yielded higher root dentin thickness than PRF. | Not specified |
| 21 | Kalash et al., 2017 [44] | Immediate implant placement and filling of peri-implant gap with xenograft or PRF–xenograft mixture. | 18 | Days 2, 7 and 14; 3, 6 and 9 months | The A-PRF and xenograft mixture positively affected soft tissue healing and bone regeneration. Improvement in implant stability was noted, with statistically significant difference. | None |
| 22 | Lavagen et al., 2021 [45] | The usage of A-PRF in the treatment of alveolar cleft with iliac bone graft. In the study, the authors evaluated the efficiency of using A-PRF by comparing the volumes of newly formed bone after a bone graft combining autogenous iliac crest bone with either PRF or A-PRF. | 24 | 6 months | In groups with A-PRF placement, bone regeneration was more effective. | Not specified |
| 23 | Nowak et al., 2021 [46] | The effect of A-PRF application during surgical extraction of third molars on healing and the concentration of C-reactive protein. | 60 | 7 days | A faster decrease in C-reactive protein levels was shown in patients who used A-PRF after third molar extraction. A-PRF accelerated healing and reduced the occurrence of alveolar osteitis. | Not specified |
| 24 | Öngöz Dede et al., 2023 [47] | The clinical effects of concentrated growth factor and A-PRF applied together using the CAF technique in the treatment of type I multiple gingival recessions. | 16 | 6 months | Significant improvements were determined in the clinical attachment level, vertical gingival recession, horizontal gingival recession, gingival thickness, width of keratinized gingiva, percentages of the mean and complete root coverage at 6 months in the CAF + A-PRF group. Mean root coverage was the best in the CAF + A-PRF group. | Not specified |
| 25 | Pereira et al., 2023 [48] | The effects of A-PRF+ on the healing of upper third molar post-extraction sockets. | 16 | 90 days | There were no clinical differences regarding healing in any control follow-up. | Not specified |

**Table 1.** *Cont.*

| No | References | Aim of the Study | Number of Patients | Follow-Up | Results | Complications |
|---|---|---|---|---|---|---|
| 26 | Praganta et al., 2024 [49] | The application of A-PRF and gelatin dressing in extraction sockets following mandibular wisdom teeth removal and the influence on postoperative pain and swelling. | 87 | 7 days | A-PRF placement in third molar sockets did not reduce postoperative pain and swelling compared to gelatin dressing alone. | Not specified |
| 27 | Rachna M et al., 2024 [50] | The effects of application of A-PRF or A-PRF and the eggshell membrane after teeth extraction. | 20 | 3 and 6 months | In the A-PRF and eggshell membrane group, after 3 and 6 months, the bone density in the cone-beam computed tomography scan was higher than in the A-PRF only group. | Not specified |
| 28 | Şen DÖ et al., 2024 [51] | The effects of utilizing L-PRF and A-PRF as a palatal bandage following free gingival graft on patients' morbidity and oral-health-related quality of life. | 39 | 1–7 and 14 days; 1 and 6 months | The control group without growth factors had higher OHIP-14 total scores than the other groups. The PRF groups showed an improvement in the quality of life and took less painkillers. | Not specified |
| 29 | Soto-Penaloza et al., 2019 [52] | Apical root resection (3 mm) with or without the use of A-PRF during free-flap closure. | 50 | 7 days | The difference in pain was not significant between the control and test groups. Taking into account the overall improvement in the quality of life in the test group, A-PRF can be considered as a useful addition to endodontic surgical protocol, as it provides a safe and affordable alternative. | Feeling nauseous, discomfort related to prolonged bleeding and bad breath/taste |
| 30 | Sousa et al., 2020 [53] | Patching free gingival graft sites with A-PRF clot membranes and evaluating its potential in improving wound healing. | 25 | 3 months | A-PRF membranes improved the healing process (faster decrease in the wound area and epithelialization promotion) with less postoperative pain. | Hemorrhage (control and study groups), necrosis in the control group (day 7) |
| 31 | Starzyńska et al., 2021 [19] | Assessment of the influence of A-PRF on selected clinical features following surgical removal of the impacted mandibular third molars. | 100 | 14 days | A-PRF reduced the pain intensity, analgesic intake, trismus, edema, the presence of hematomas and skin warmth. | Not specified |

**Table 1.** *Cont.*

| No | References | Aim of the Study | Number of Patients | Follow-Up | Results | Complications |
|---|---|---|---|---|---|---|
| 32 | Śmieszek-Wilczewska et al., 2024 [54] | Comparing the effectiveness of PRF and the conventional method in oroantral communication repair techniques. | 22 | 14 days | Complete wound healing. | Oroantral communications treated with A-PRF resulted in fewer complications and less pain |
| 33 | Tadepalli et al., 2023 [55] | Assessment of leukocyte platelet-rich fibrin and A-PRF in combination with CAF in the treatment of gingival recession. | 30 | 6 months | Statistically significant reduction in mean recession height was observed from baseline to 6 months in the CAF + L-PRF and CAF + A-PRF groups, respectively. The mean root coverage percentage achieved at 6 months was better in the CAF + A-PRF group (81.66 ± 28.21) than in the CAF + L-PRF group. | Not specified |
| 34 | Torul et al., 2020 [56] | Mandibular third molar extraction and evaluation of the effect of connective tissue graft and A-PRF on edema, pain and trismus. | 75 | 14 days | The study showed that connective tissue graft and A-PRF did not exert any significant effects on pain, swelling and trismus. | Not specified |
| 35 | Trimmel et al., 2021 [57] | Maxillary sinus augmentation from the lateral approach with the use of serum albumin-coated bone allograft combined with A-PRF. | 26 | 3 and 6 months | Serum albumin-coated bone allograft combined with A-PRF is a suitable material for maxillary sinus augmentation, as augmented and pristine bone showed no significant difference both in histo- and micromorphometric parameters. | In three cases (two in the control group, one in the study group), small perforation was detected and fixed with A-PRF membrane |
| 36 | Yewale et al., 2021 [58] | Atraumatic tooth extraction and socket preservation with Sybograf plus or Sybograf plus/A-PRF. | 20 | 6 months | The use of A-PRF increased the effectiveness of the bone graft used in preserving vertical and horizontal dimensions. The swelling percentage in the A-PRF group was noticeably decreased, which led to less discomfort for patients. The pain levels remained equal in both study groups. | None |

**Table 1.** *Cont.*

| No | References | Aim of the Study | Number of Patients | Follow-Up | Results | Complications |
|---|---|---|---|---|---|---|
| 37 | Yüce and Kömerik, 2019 [59] | Managing alveolar osteitis after third molar extraction using A-PRF. | 40 | 1, 3, 7 and 15 days; 1, 2 and 3 months | The use of A-PRF compared to the control group lowered the Visual Analog Scale pain score, which resulted in less analgesics taken. Statistical epithelial healing rates were faster in the A-PRF group. The gray level pixel values comparison showed improved hard tissue healing in the A-PRF group. | Not specified |
| 38 | Zahid and Nadershah, 2019 [60] | Assessing the impact on third molar extraction with the use of A-PRF as a regenerative bio-material. | 10 | 1 and 3 months | A-PRF decreased postoperative pain and swelling. A-PRF provided slight but not significant advantages in terms of probing depth reduction, recession coverage and clinical attachment level gain compared to control. | Not specified |

A-PRF—advanced platelet-rich fibrin; BCP—biphasic calcium phosphate; CAF—coronally advanced flap; CAL—clinical attachment level; DBBM—deproteinized bovine bone mineral; EMD—enamel matrix derivative; FDBA—freeze-dried bone allografts; FMBS—full mouth bleeding score; HEM—hemostatic plug; L-PRF—leukocyte platelet-rich fibrin; OHIP-14—Oral Health Impact Profile.

Most often, A-PRF is compared to L-PRF or blood cloth. A radiological analysis performed by Castro et al. [30] showed that A-PRF presented better results than blood cloth alone in all analyzed points (horizontal, buccal, palatal resorptions, ridge width changes, vertical resorption and socket fill) but fared very similar to L-PRF in terms of alveolar ridge preservation. A morphometric bone analysis (histological, 2D and 3D micro-computed tomography (micro-CT)) presented similar results between the study groups, with both outperforming the blood cloth control group. Although both PRFs hindered bone resorption, they could not counteract it completely.

Two randomized controlled trials (RCTs) conducted by Clark et al. [32] and Ivanova et al. [39] also include the analysis of A-PRF behavior when combined with FDBA in alveolar ridge preservation.

In the operating procedure described by Clark [32], mucoperiosteal flap elevation was not performed; single-root teeth were extracted; and non-steroidal anti-inflammatory drugs (NSAIDs; 600 mg ibuprofen) and a 0.12% chlorhexidine mouth rinse were administered at the beginning of the surgery. Conversely, Ivanova [36] elevated the flap from both the palatal and buccal sides, extracted multiple-root teeth and administered a 0.12% mouth rinse twice daily for a period of two weeks following the surgical procedure. Additionally, they administered an antibiotic (1000 mg amoxicillin) and a non-steroidal anti-inflammatory drug (nimesulide 100 mg).

For the purpose of measuring the dimensions of the operating area, Ivanova et al. [39] employed the use of a Trios intraoral scanner for the analysis of a virtual model, whereas Clark et al. [32] utilized a more conventional methodology, namely the fabrication of a stent from light-cured resin derived from an alginate impression, which was then measured with a periodontal probe for height and calipers for width. In both cases, a trepanning burr was used to harvest the bone core. The primary outcome was assessed in both randomized controlled trials (RCTs) by means of a comparison of the vertical and horizontal dimensions of the alveolar ridge, while the secondary and tertiary outcomes were evaluated through histomorphometric and micro-computed tomography (micro-CT) analyses.

Clark et al. [32] observed that the groups using A-PRF and FDBA demonstrated lower alveolar ridge height reduction compared to the blood cloth group. Histomorphometric analysis showed that A-PRF and FDBA + A-PRF allowed the formation of a denser trabecular structure. In terms of augmenting material integration, A-PRF enhanced the integration of FDBA by decreasing the amount of residual graft material and showed significantly higher maturation of the structure compared to FDBA alone. Bone vitality was greatest in the A-PRF group, but mineral bone density was best in the FDBA group; both results showed statistical significance. The study conducted by Ivanova et al. [23] presented similar results regarding the superiority of A-PRF and FDBA over blood cloth regarding alveolar ridge dimension preservation, but in the case of vital bone formation, A-PRF did not perform better than FDBA.

In contrast, Pereira et al. investigated the effect of A-PRF on socket healing after extraction of upper wisdom teeth. A clinical and cone-beam computed tomography (CBCT) study was conducted, and no significant benefit was found between A-PRF application and treatment without the use of growth factors [48].

#### 3.2.2. A-PRF Effect on Postoperative Pain, Swelling and Trismus

A meta-analysis of nine studies revealed that A-PRF administration was demonstrated to effectively reduce postoperative pain in seven studies [19,24,31,37,46,49,51,56,60].

Caymaz et al. [31] compared A-PRF to L-PRF in terms of pain, swelling and trismus after third molar extractions. Apart from the VAS pain score, in the study, pain was also measured by analgesic drug usage. The most noticeable difference was visible in the first 3 postoperative days. Compared with L-PRF, A-PRF presented much lower VAS pain scores in the first 3 postoperative days, which slowly equaled the L-PRF group by day 7. The difference in the number of analgesics used by the A-PRF group was noticeably lower from days 2 to 6. In terms of swelling and trismus, there were no significant differences between the two groups.

Zahid and Nadershah [60] performed third molar extractions with the use of A-PRF as a test group or with blood cloth only as a control group. The A-PRF group showed decreased pain and swelling compared to the control group over 7 postoperative day periods. Moreover, the healing aspects were checked (pocket depth, gingival recession, clinical attachment level), but no significant advantages were detected compared to the A-PRF group.

In the study by Nowak et al. [46], the impact of A-PRF application on the surgical removal of third molars was examined, with a particular emphasis on its influence on healing and the concentration of C-reactive protein. A more rapid decline in C-reactive protein levels was observed in subjects who underwent third molar extraction and subsequently utilized A-PRF. The application of A-PRF resulted in accelerated healing and a reduction in the incidence of alveolar osteitis.

Starzyńska et al. [19] assessed the influence of A-PRF on selected clinical features following surgical removal of the impacted mandibular third molars. A-PRF reduced the pain intensity, analgesic intake, trismus, edema, the presence of hematomas and skin warmth.

In contrast to the studies above, Torul et al. [56] did not detect significant advantages of using A-PRF in the procedure of lower third molar extraction. The control group (blood cloth) was compared with CGF and A-PRF groups in terms of pain, swelling and trismus. The only significant result regarding A-PRF usage was detected in terms of swelling on the seventh postoperative day compared to the CGF group (the Tragus–Pogonion measurement). In all categories, A-PRF exhibited a similar outcome to that of the control group, demonstrating minimal improvement. Additionally, Praganta et al. [49] did not identify a notable distinction in the reduction in pain and swelling between the A-PRF and gelatin sponge groups following the extraction of wisdom teeth.

The study conducted by Hartlev et al. [37] researched A-PRF usability in pain management in the surgical procedure of lateral ridge augmentation following mandibular ramus block harvesting. The test group, which utilized A-PRF, demonstrated superior results in terms of VAS pain scoring compared to the control group, which employed a blood cloth. Although throughout the seven measured days, pain decrease was only significant in the first postoperative day, the A-PRF group showed lowered VAS pain score up to the fifth day. Taking into account the fact that the pain associated with the surgical protocol used is usually low according to the authors, decreasing the pain levels further can still improve the quality of life of patients.

In the case of the treatment of recession with the pin hole method and A-PRF simultaneously in the study by Al-Barakani et al. [24], a reduction in the level of postoperative pain was observed.

In the study by Şen DÖ et al. [51], the effects of L-PRF and A-PRF as a palatal bandage following free gingival graft on patients' morbidity and oral-health-related quality of life were examined. The control group without growth factors had higher OHIP-14 total scores than the other groups. The PRF groups showed an improvement in the quality of life and took less painkillers.

#### 3.2.3. A-PRF Use in Implantology

The field of implantology is dedicated to identifying the most effective solutions and alternatives to address the challenges encountered by patients. A review of the literature revealed three randomized controlled trials investigating the impact of A-PRFs in surgical protocols.

The studies conducted by Kalash et al. [44] and Alhaj et al. [25] focused on the effect of A-PRF in immediate implantation when combined with grafting material. Both pieces of research indicated that A-PRF addition improved the clinical outcomes. The probing depth and implant stability checked by Kalash et al. [44] outperformed the control group, showing fewer variations in the results at follow-up points. The addition of A-PRF to the surgical protocol was found to enhance the marginal bone height and bone density in both studies. The study by Alhaj et al. [25] demonstrated a statistically significant difference between the groups at the final follow-up.

The third study concentrated on the comparison of A-PRF and FGG, with an evaluation of their potential for improving the keratinized tissue around implants. Alsahli et al. [26] found that there were no significant differences between the two groups, which indicated that A-PRF performed similarly to FGG. The use of A-PRF was shown to lower the postoperative pain up to the sixth day. In consideration of the absence of a second surgical site, A-PRF emerges as a compelling alternative to the established protocol. The drawback is that the thickness of the keratinized tissue in the A-PRF group showed a gradual decrease over time.

#### 3.2.4. A-PRF in Hard Tissue Healing

To date, six randomized controlled studies have been conducted on the use of A-PRF alone [27,38,58] or in conjunction with other PRF types [33,45] or membranes [50]. A-PRF was highlighted to have a positive effect on grafting material integration in three of these studies [35,37,38]. Furthermore, the A-PRF groups indicated enhanced biometric bone quality with lower grafting material resorption. This allows for maintaining the alveolar ridge dimensions more predictably, which leads to more predictable future implant placements. A study by Dayashankara Rao et al. [33] also focused on the periodontal status and pocket depth in recipient places. The test group performed better than the control, showing improvement in both aspects with lower mobility scores. Yewale et al. [51] checked the pain and swelling of the test and control groups. While the pain levels remained similar between the two groups, the swelling decreased noticeably in the test group, improving patient postoperative comfort. Angelo et al. [27] used A-PRF as the membrane for a biomaterial in piezotome-enhanced subperiosteal tunnel technique (PeSPTT). The results demonstrated that the use of A-PRF resulted in more consistent outcomes, with enhanced bone formation quality for implant placement.

However, the study conducted by Hartlev et al. [38] presents a contrasting approach, directly comparing A-PRF membranes covering autograft material with autografts combined with DBBM and a collagen membrane. The test and control groups showed no significant differences in terms of vital and non-vital bone formation, the amount of new blood vessels formed and the improvement in soft tissues.

In a previous study, Lavagen and colleagues [45] demonstrated the efficacy of A-PRF in the treatment of alveolar clefts with iliac bone grafts. In the study, the authors evaluated the efficiency of using A-PRF by comparing the volumes of newly formed bone after a bone graft combining autogenous iliac crest bone with either PRF or A-PRF. In groups with A-PRF placement, bone regeneration was more effective.

In the study conducted by Rachna et al. [50], the impact of applying A-PRF or A-PRF combined with eggshell membrane following tooth extraction was investigated. In the A-PRF and eggshell membrane group, after 3 and 6 months, the bone density in the CBCT scan was higher than in the A-PRF group.

#### 3.2.5. A-PRF in Soft Tissue Healing

The present study analyzed seven studies [24,28,29,40,44,47,53] that were based on research on palatal FGG surgical procedures. A-PRF was made the test group for patching the donor site. The clinical outcomes take into account direct soft tissue changes (color changes, contour changes, texture, epithelialization, wound area reduction) and postoperative pain (VAS score). In terms of color, contour and texture, in the controlled trial conducted by Bahmanamm [28], A-PRF showed better results than the control group. The wound margin analysis showed a consistently better score for the A-PRF group. Moreover, A-PRF promoted more efficient epithelialization of donor sites and enhanced wound healing. The effect was most profound from day 7 to 30. In that period, the percentage wound area reduction in the study by Sousa et al. [53] showed a statistically significant (7th and

30th day) or close to significant (14th day) result. Regarding pain perception, the VAS score measured in both studies [40,44] showed reductions in the A-PRF group, with statistical significance in Bahmanamm's trial [28].

In their study, Öngöz Dede et al. [47] analyzed the effects of concentrated growth factors in combination with A-PRF used during coronally advanced flap in the treatment of multiple gingival recessions (type 1 recessions). In the case of using platelet-rich preparations in the treatment of recession, significant improvements were determined in the clinical attachment level, vertical gingival recession, horizontal gingival recession, gingival thickness, width of keratinized gingiva, percentages of the mean and complete root coverage compared to a control group that did not receive CGF and A-PRF.

In the study conducted by Al-Barakani et al. [24], the treatment of recession (types I and II, as defined by the Miller classification) using A-PRF in the pinhole surgical technique (PST) was demonstrated to be more effective than the application of a resorbable collagen membrane (RCM) in PST. However, both methods were ultimately deemed unsatisfactory.

In a further study, Barakat et al. [29] examined the clinical effects of A-PRF and connective tissue grafts (CTGs) in the treatment of gingival black triangle (GBT) using Han and Takei's method. The results of the study indicated that both A-PRF and CTG yielded comparable outcomes when used in conjunction with Han and Takei's techniques.

#### 3.2.6. A-PRF Effect on Hemostasis

The studies conducted by Brancaccio et al. [2] and Giudice et al. [33] gathered a total of 142 patients taking antiplatelet drugs. Both studies used the direct comparison of blood cloth (control), hemostatic plug (HEM), A-PRF and L-PRF (test groups). Each patient had four non-adjacent teeth removed, without flap elevation. Based on the trials, A-PRF exerted the best hemostatic effect. When postoperative bleeding was assessed 30 min after the extractions, A-PRF demonstrated superior performance to the control group, with statistical significance [37]. Branaccio et al. [2] obtained similar results on the broader patient groups but with both A-PRF and L-PRF significantly outperforming the control group. Furthermore, A-PRF showed statistical significance in bleeding reduction compared to HEM. Further evaluation of wound healing showed that A-PRF and L-PRF promoted wound closure more effectively than HEM and the control in both studies [2,36]. Regarding patient preference, Giudice et al. [36] surveyed patients one and two weeks after the extractions. After one week, most patients selected A-PRF, with control sites being the second choice. On the second follow-up, A-PRF and L-PRF were ex aequo the most often selected options.

#### 3.2.7. A-PRF in Maxillary Sinus Augmentation

There have been two randomized controlled trials identified regarding A-PRF use in maxillary sinus lift. Trimmel et al. [57] followed the augmentation by placing the implants in the augmented sites, measuring the implant stability quotient (ISQ) throughout the osseointegration process. The control group showed a significantly better ISQ at weeks 6 and 8. During implant placement, bone samples were collected and analyzed. The micromorphometry results favored the A-PRF group, as it presented significantly better results in terms of bone surface/bone volume ratio, bone surface density, trabecular thickness and connectivity. Histomorphometry using hematoxylin–eosin staining did not present a difference between the two groups.

In contrast, Dragonas et al. [34] compared A-PRF and plasma rich in growth factors combined with deproteinized bovine bone mineral during sinus lift augmentation. The mean percentage of mineralized bone after the healing period was higher in the group with growth factors, but there was no statistical significance in samples without growth factors. Adding A-PRF and PRGF to DBBM did not improve new bone formation in the sinus lift procedure. Neither platelet-rich preparation was better than the other in any of the parameters studied.

#### 3.2.8. A-PRF in Intrabony Defect Management

Additionally, A-PRF is emerging as a material with valuable applications in periodontology. The study by Ghonima et al. [35] compared A-PRF + BCP with BCP + saline. The test group did not show statistically significant improvements over the control group, though it must be noted that the outcomes were slightly more favorable toward A-PRF in terms of PD reduction and CAL gain. Both groups also showed a significant reduction in the plaque index at 9-month follow-up. On the other hand, Csifó-Nagy et al. [21] compared A-PRF with EMD. The A-PRF test group performed similarly to the EMD group, with no statistical differences in terms of PD, GR, CAL and bone sounding changes. Moreover, both materials used in the procedure lowered the full mouth bleeding score at a 6-month follow-up.

#### 3.2.9. A-PRF Use in Alveolar Osteitis (Dry Socket)

The research conducted by Yüce et al. [59] focused on the pain levels and soft and hard tissue differences in patients suffering from alveolar osteitis. The debridement and irrigation were accompanied by A-PRF in the test group. In terms of pain reduction, researchers found that the pain decreased rapidly and continually on days 1, 3, 5 and 7 in conjunction with analgesic usage. Compared with control, the results were statistically significant. Looking into soft tissue changes, the test group presented significantly higher epithelium healing rates at all times. The degree of hard tissue healing was estimated by calculating the gray level pixel count. The measurements taken at second and third months showed statistical significance in average pixel values for the test group.

#### 3.2.10. A-PRF in Endodontic Surgery

There have been two studies that examined A-PRF usage in endodontics [40,50]. A-PRF was introduced to the apical root resection protocol with paramarginal mucoperiosteal flap release. Patients were only selected if they required endodontic treatment of the maxillary second premolar and had lesions measuring 6–12 mm (cone-beam computed tomography measurement). In terms of pain reduction, A-PRF lowered the VAS score, but with no statistical significance. The same was true for analgesic usage. A-PRF also fared better than control in mouth opening and chewing functions, sleep impairment, inflammation of the operation site and discomfort in that area [50].

In the study by Jayadevan et al. [40], A-PRF and PRF were used as a scaffold in the regenerative endodontic treatment (RET) of traumatized immature non-vital teeth. Immature teeth have narrow canal walls. A-PRF yielded higher root dentin thickness than PRF.

#### 3.2.11. A-PRF in Treatment of Oroantral Communication

In the study by Śmieszek-Wilczewska et al. [54], oroantral communication treated with A-PRF resulted in fewer complications and less pain. In addition, the use of A-PRF alone as clots resulted in complete healing and closure of the connection. Additional methods, such as regional flaps, were not necessary.

### 3.3. *Quality Assessment*

The risk of bias assessment using RoS 2 is described in Figure 3.

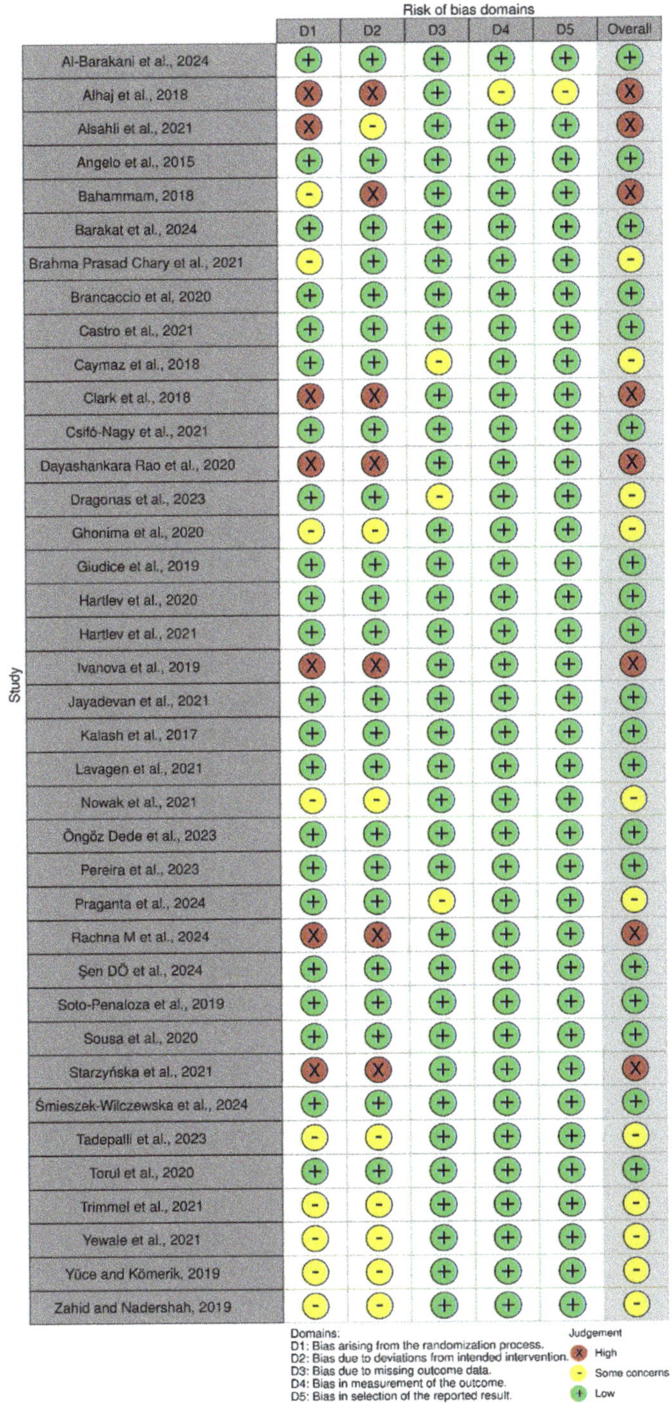


**Figure 3.** The risk of bias assessment using RoS 2 [2,19–21,24–40,44–60].

## 4. Discussion

A-PRF is still a novel material, related to the original PRF; nonetheless, further research must be conducted on its applications. The existing randomized controlled trials proved to be a scarce source of data regarding the sole use of A-PRF in dental surgical procedures. The variety of different combinations of bio-materials used with advanced platelet-rich fibrin often led to slightly altered results between the trials regarding specific usage. The vast majority of studies limited themselves to small patient cohorts, rarely exceeding 40 patients in the test groups. The follow-up periods were also a variable factor between the studies, depending mainly on the procedure performed but also on the protocol followed by the surgeons. In the presented studies, the cohorts were divided mostly equally, with few exceptions [31,32,37,38,53]. Despite these differences, the results demonstrated that A-PRF has a vast spectrum of applications and provides a valuable supplementary benefit when integrated into the surgical protocol. Studies that employed A-PRF as the sole comparison standard primarily investigated its efficacy in preserving alveolar ridge dimensions, accelerating healing, promoting epithelialization, enhancing hemostasis, alleviating pain, reducing trismus, minimizing swelling and mitigating the complications associated with tooth extractions, including mouth opening limitations [2,19,21,24–32,35–39,46,51–54,59,60]. The results presented the use of A-PRF as beneficial, even though A-PRF requires additional steps and machinery, prolonging the overall surgery time and requiring blood donation from the patient.

Regarding the combination of A-PRF with other bio-materials, 13 out of 36 analyzed randomized controlled trials used it in the research. It was most popular in works examining the improvement of new bone formation in conjunction with bone grafting materials [20,25,27,32–39,44,45,57,58]. The evidence of combining A-PRF with grafting materials seems to provide significantly better results, although not all studies obtained favorable results [37]. The main benefit seems to rely on A-PRF providing the essential growth factors quickly, enhancing the bio-material turnover. In two of the studies [54,57], A-PRF also seemed suitable in the event of closing perforation in maxillary sinus mucosa. The positive effect of A-PRF in terms of synergy with bio-materials and improvement of surgery outcomes allows for favorable implantation conditions and more optimized resource usage.

A comparison of the results with other systematic reviews on the topic of A-PRF is not currently possible, as to our knowledge, this is the first review to provide a comprehensive overview of the current A-PRF use cases, which are supported by randomized controlled trials. With the growing interest in natural materials and technological advancements, the future use of A-PRF is only going to expand.

Even though the evidence is of mid-to-high quality, the results still have to be interpreted with caution. Most of the studies specified different approaches to the specific surgeries performed. The variety of bio-materials employed, including autografts, xenografts and enamel matrix derivatives, along with the diverse range of postoperative drug regimens and follow-up protocols, make it challenging to predict whether a consistent outcome can be achieved through the typical surgical approach. Although most studies blinded the participants and randomized the cohorts using special programs to achieve the best randomization, there is always some bias included. Moreover, some of the studies included surgeries performed by more than one surgeon, which can alter the results ever so slightly. The focus on the specific parameters examined can also overshadow other variables, which can influence the results. Although no complications were noted, and patients attended the follow-ups in most studies, there is no guarantee that, on a larger scale, the outcome would be the same. Language limitation was also present in this study, as only publications in English were chosen for conducting this systematic review.

## 5. Conclusions

The utilization of A-PRF is broadening rapidly. There is growing evidence that advanced platelet-rich fibrin has a positive effect on the surgical procedures mentioned. The randomized controlled trials presented mostly favorable results, though it must be noted that such results were observed on a smaller scale. The rapid growth of new technologies, mainly implantology, will accompany the wider and further adoption of A-PRF. Due to its biological properties, A-PRF could be considered a reliable addition to the surgical protocols, both alone and as an additive to bio-materials, with the advantages of healing improvement, pain relief, soft tissue management and bone preservation, as well as graft integration. Nevertheless, further well-designed randomized controlled trials are required for each use case. Ideally, these would include larger patient cohorts, additional research personnel blinded to the treatment allocation and long-term follow up periods. This would allow a determination of the long-term clinical implications and recommendations for clinical practice.

**Author Contributions:** Conceptualization, P.A. and M.C.; methodology, M.C. and P.A.; software, M.C. and P.A.; validation, M.C. and P.A.; formal analysis, M.C. and P.A.; investigation, M.C. and P.A.; resources, M.C. and P.A.; data curation, M.C. and P.A.; writing—original draft preparation, M.C. and P.A.; writing—review and editing, M.C., P.A. and A.P.; visualization, P.A. and M.C.; supervision, M.C., P.A. and A.P.; project administration, A.P. All authors have read and agreed to the published version of the manuscript.

**Funding:** This research received no external funding.

**Institutional Review Board Statement:** Not applicable.

**Informed Consent Statement:** Not applicable.

**Data Availability Statement:** The original contributions presented in the study are included in the article, further inquiries can be directed to the corresponding author.

**Conflicts of Interest:** The authors declare no conflicts of interest.

## Abbreviations

| | |
|---|---|
| AFG | autologous fibrin glue |
| A-PRF | advanced platelet-rich fibrin |
| A-PRF+ | advanced platelet-rich fibrin plus |
| BCP | biphasic calcium phosphate |
| BMPs | bone morphogenic factors |
| BMP-2 | bone morphogenic factor 2 |
| $Ca^{2+}$ | calcium ion |
| CAL | clinical attachment level |
| CBCT | cone-beam computed tomography |
| CGF | concentrated growth factor |
| CRC | complete root coverage |
| CRP | C-reactive protein |
| CTG | connective tissue graft |
| DBBM | demineralized bovine bone mineral |
| EMD | enamel matrix derivative |
| FDBA | freeze-dried bone allografts |
| FGF | fibroblast growth factor |
| FGG | free gingival graft |
| FMBS | full mouth bleeding score |
| GBT | gingival black triangle |
| GR | gingival recession |
| GT | gingival thickness |
| HEM | hemostatic plug |
| HGR | horizontal gingival recession |
| IL-1β | interleukin 1β |
| IL-4 | interleukin IL-4 |
| IL-6 | interleukin IL-6 |
| I-PRF | injectable platelet-rich fibrin |

| | |
|---|---|
| ISQ | implant stability quotient |
| KGW | width of keratinized gingiva |
| L-PRF | leukocyte–platelet-rich fibrin |
| micro-CT | micro-computed tomography |
| MMPs | matrix metalloproteinases |
| MMP-9 | matrix metalloproteinases 9 |
| MRC | percentages of the mean |
| NSAID | non-steroidal anti-inflammatory drugs |
| OHIP-14 | Oral Health Impact Profile |
| PD | pocket depth |
| PDGF | platelet-derived growth factor |
| PeSPTT | piezotome-enhanced subperiosteal tunnel technique |
| PICO | population, intervention, comparison, outcome |
| PRGF | plasma rich in growth factors |
| PRISMA | Preferred Reporting Items for Systematic Reviews and Meta-Analyses |
| PROSPERO | Prospective Register of Systematic Reviews |
| PRF | platelet-rich fibrin |
| PRGF | plasma rich in growth factors |
| PRP | platelet-rich plasma |
| PST | pin hole surgical technique |
| RCM | resorbable collagen membrane |
| RCTs | randomized clinical trials |
| RET | regenerative endodontic treatment |
| SACBA | serum albumin-coated bone allograft |
| T-PRF | titanium platelet-rich fibrin |
| TGF | transforming growth factor |
| TNFα | tumor necrosis factor α |
| TGFα | tumor necrosis factor α |
| TGFβ | tumor necrosis factor β |
| VAS | Visual Analog Scale |
| VEGF | vascular-endothelial growth factor |
| VGR | vertical gingival recession |

## References


1. Dohan, D.M.; Choukroun, J.; Diss, A.; Dohan, S.L.; Dohan, A.J.J.; Mouhyi, J.; Gogly, B. Platelet-rich fibrin (PRF): A secondgeneration platelet concentrate. Part I: Technological concepts and evolution. *Oral Surg. Oral Med. Oral Pathol. Oral Radiol. Endod.* **2006**, *10*, 37–44. [CrossRef]
2. Brancaccio, Y.; Antonelli, A.; Barone, S.; Bennardo, F.; Fortunato, L.; Giudice, A. Evaluation of local hemostatic efficacy after dental extractions in patients taking antiplatelet drugs: A randomized clinical trial. *Clin. Oral Investig.* **2021**, *25*, 115–1167. [CrossRef]
3. Choukroun, J.; Diss, A.; Simonpieri, A.; Girard, M.O.; Schoeffler, C.; Dohan, S.L.; Dohan, A.J.; Mouhyi, J.; Dohan, D.M. Platelet-rich fibrin (PRF): A second-generation platelet concentrate. Part IV: Clinical effects on tissue healing. *Oral Surg. Oral Med. Oral Pathol. Oral Radiol. Endod.* **2006**, *101*, e56–e60. [CrossRef] [PubMed]
4. Choukroun, J.; Diss, A.; Simonpieri, A.; Girard, M.O.; Schoeffler, C.; Dohan, S.L.; Dohan, A.J.; Mouhyi, J.; Dohan, D.M. Platelet-rich fibrin (PRF): A second-generation platelet concentrate. Part V: Histologic evaluations of PRF effects on bone allograft maturation in sinus lift. *Oral Surg. Oral Med. Oral Pathol. Oral Radiol. Endod.* **2006**, *101*, 299–303. [CrossRef] [PubMed]
5. Ghanaati, S.; Booms, P.; Orlowska, A.; Kubesch, A.; Lorenz, J.; Rutkowski, J.; Landes, C.; Sader, R.; Kirkpatrick, C.; Choukroun, J. Advanced platelet-rich fibrin: A new concept for cell-based tissue engineering by means of inflammatory cells. *J. Oral Implantol.* **2014**, *40*, 679–689. [CrossRef] [PubMed]
6. Choukroun, J. Advanced PRF, and i-PRF: Platalet concentrates or blood concentrates? *J. Periodont. Med. Clin. Pract.* **2014**, *1*, 3.
7. Upadhayaya, V.; Arora, A.; Goyal, A. Bioactive Platelet Aggregates: Prp, Prgf, Prf, Cgf And Sticky Bone. *IOSR J. Dent. Med. Sci.* **2017**, *16*, 5–11. [CrossRef]
8. Shirbhate, U.; Bajaj, P. Third-Generation Platelet Concentrates in Periodontal Regeneration: Gaining Ground in the Field of Regeneration. *Cureus* **2022**, *14*, e28072. [CrossRef] [PubMed]
9. Illmilda; Asrianti, D.; Margono, A.; Julianto, I.; Wardoyo, M.P. Advanced Platelet Rich Fibrin (A-PRF) supplemented culture medium for human dental pulp stem cell proliferation. *J. Int. Dent. Med. Res.* **2019**, *12*, 396–400.
10. Kobayashi, E.; Flückiger, L.; Fujioka-Kobayashi, M.; Sawada, K.; Sculean, A.; Schaller, B.; Miron, R.J. Comparative release of growth factors from PRP, PRF, and advanced-PRF. *Clin. Oral Investig.* **2016**, *20*, 2353–2360. [CrossRef]

11. Dohan Ehrenfest, D.M.; Diss, A.; Odin, G.; Doglioli, P.; Hippolyte, M.P.; Charrier, J.B. In vitro effects of Choukroun's PRF (platelet-rich fibrin) on human gingival fibroblasts, dermal prekeratinocytes, preadipocytes, and maxillofacial osteoblasts in primary cultures. *Oral Surg. Oral Med. Oral Pathol. Oral Radiol. Endod.* **2009**, *108*, 341–352. [CrossRef]
12. Gupta, A.K.; Cole, J.; Deutsch, D.P.; Everts, P.A.; Niedbalski, R.P.; Panchaprateep, R.; Rinaldi, F.; Rose, P.T.; Sinclair, R.; Vogel, J.E.; et al. Platelet-Rich Plasma as a Treatment for Androgenetic Alopecia. *Dermatol. Surg.* **2019**, *45*, 1262–1273. [CrossRef] [PubMed]
13. Everts, P.A.M.; Knape, J.T.; Weibrich, G.; Schönberger, J.P.; Hoffmann, J.; Overdevest, E.P.; Box, H.A.; van Zundert, A. Platelet-rich plasma and platelet gel: A review. *J. Extracorpor. Technol.* **2006**, *38*, 174–187. [CrossRef]
14. Kargarpour, Z.; Nasirzade, J.; Panahipour, L.; Mitulović, G.; Miron, R.J.; Gruber, R. Platelet-Rich Fibrin Increases BMP2 Expression in Oral Fibroblasts via Activation of TGF-β Signaling. *Int. J. Mol. Sci.* **2021**, *22*, 7935. [CrossRef] [PubMed]
15. Steller, D.; Herbst, N.; Pries, R.; Juhl, D.; Hakim, S.G. Impact of incubation method on the release of growth factors in non-Ca2+-activated PRP, Ca2+-activated PRP, PRF and A-PRF. *J. Craniomaxillofac. Surg.* **2019**, *47*, 365–372. [CrossRef]
16. Miron, R.J.; Fujioka-Kobayashi, M.; Sculean, A.; Zhang, Y. Optimization of platelet-rich fibrin. *Periodontology 2000* **2024**, *94*, 79–91. [CrossRef]
17. Miron, R.J. *Understand Platelet Rich Fibrin*, 1st ed.; Quintessence Publishing Co, Inc.: Batavia, IL, USA, 2021.
18. Machut, K.; Zoltowska, A.; Pawlowska, E.; Derwich, M. Plasma Rich in Growth Factors in the Treatment of Endodontic Periapical Lesions in Adult Patients: Case Reports. *Int. J. Mol. Sci.* **2021**, *22*, 9458. [CrossRef]
19. Starzyńska, A.; Kaczoruk-Wieremczuk, M.; Lopez, M.A.; Passarelli, P.C.; Adamska, P. The Growth Factors in Advanced Platelet-Rich Fibrin (A-PRF) Reduce Postoperative Complications after Mandibular Third Molar Odontectomy. *Int. J. Environ. Res. Public Health* **2021**, *18*, 13343. [CrossRef]
20. Brahma Prasad Chary, N.O.; Raju, M.S.; Suresh Sajjan, M.C.; Gottumukkala, S.N.; Manyam, R. Comparison of quality of bone and insertion torque values of early implants placed at 6 and 8 weeks in sockets preserved with advanced platelet-rich fibrin: A randomized controlled trial. *J. Indian Prosthodont. Soc.* **2021**, *21*, 366–374. [CrossRef] [PubMed]
21. Csifó-Nagy, B.K.; Sólyom, E.; Bognár, V.L.; Nevelits, A.; Dőri, F. Efficacy of a new-generation platelet-rich fibrin in the treatment of periodontal intrabony defects: A randomized clinical trial. *BMC Oral Health* **2021**, *21*, 580. [CrossRef]
22. Liu, Y.H.; To, M.; Okudera, T.; Wada-Takahashi, S.; Takahashi, S.S.; Su, C.Y.; Matsuo, M. Advanced platelet-rich fibrin (A-PRF) has an impact on the initial healing of gingival regeneration after tooth extraction. *J. Oral Biosci.* **2022**, *64*, 141–147. [CrossRef] [PubMed]
23. Bao, M.; Du, G.; Zhang, Y.; Ma, P.; Cao, Y.; Li, C. Application of Platelet-Rich Fibrin Derivatives for Mandibular Third Molar Extraction Related Post-Operative Sequelae: A Systematic Review and Network Meta-Analysis. *J. Oral Maxillofac. Surg.* **2021**, *79*, 2421–2432. [CrossRef]
24. Al-Barakani, M.S.; Al-Kadasi, B.; Al-Hajri, M.; Elayah, S.A. A comparative study of the effects of advanced platelet-rich fibrin and resorbable collagen membrane in the treatment of gingival recession: A split-mouth, randomized clinical trial. *Head Face Med.* **2024**, *20*, 41. [CrossRef] [PubMed]
25. Alhaj, F.; Shokry, M.; Attia, N. The efficiency of using advanced platelet rich fibrin–Autogenous bone graft mixture around immediately placed dental implants in mandibular molar region: (Randomized controlled clinical trial). *Egypt. Dent. J.* **2018**, *64*, 2023–2035. [CrossRef]
26. Alsahli, J.; Kasem, T.; Alkhouli, M. Evaluation of Apically Positioned flap with A_PRF Vs. Free Gingival Grafts to Enhance the Keratinized Tissue Around Dental Implants: A Randomized Controlled Clinical Split Mouth Trial. *Int. J. Dent. Oral Sci.* **2021**, *08*, 1844–1850. [CrossRef]
27. Angelo, T.; Marcel, W.; Andreas, K.; Izabela, S. Biomechanical Stability of Dental Implants in Augmented Maxillary Sites: Results of a Randomized Clinical Study with Four Different Biomaterials and PRF and a Biological View on Guided Bone Regeneration. *BioMed Res. Int.* **2015**, *2015*, 850340. [CrossRef]
28. Bahammam, M.A. Effect of platelet-rich fibrin palatal bandage on pain scores and wound healing after free gingival graft: A randomized controlled clinical trial. *Clin. Oral Investig.* **2018**, *22*, 3179–3188. [CrossRef] [PubMed]
29. Barakat, S.O.; Tawfik, O.K.; Kholy, S.E.; ElNahass, H. Evaluation of advanced platelet-rich fibrin compared to subepithelial connective tissue graft in the surgical management of interdental papilla recession: A randomized controlled trial. *Clin. Oral Investig.* **2024**, *28*, 87. [CrossRef] [PubMed]
30. Castro, A.B.; Van Dessel, J.; Temmerman, A.; Jacobs, R.; Quirynen, M. Effect of different platelet-rich fibrin matrices for ridge preservation in multiple tooth extractions: A split-mouth randomized controlled clinical trial. *J. Clin. Periodontol.* **2021**, *48*, 984–995. [CrossRef]
31. Caymaz, M.G.; Uyanik, L.O. Comparison of the effect of advanced platelet-rich fibrin and leukocyte- and platelet-rich fibrin on outcomes after removal of impacted mandibular third molar: A randomized split-mouth study. *Niger. J. Clin. Pract.* **2019**, *22*, 546–552. [CrossRef]
32. Clark, D.; Rajendran, Y.; Paydar, S.; Ho, S.; Cox, D.; Ryder, M.; Dollard, J.; Kao, R.T. Advanced platelet-rich fibrin and freeze-dried bone allograft for ridge preservation: A randomized controlled clinical trial. *J. Periodontol.* **2018**, *89*, 379–387. [CrossRef]
33. Dayashankara Rao, J.K.; Bhatnagar, A.; Pandey, R.; Arya, V.; Arora, G.; Kumar, J.; Bootwala, F.; Devi, W.N. A comparative evaluation of iliac crest bone graft with and without injectable and advanced platelet rich fibrin in secondary alveolar bone grafting for cleft alveolus in unilateral cleft lip and palate patients: A randomized prospective study. *J. Stomatol. Oral Maxillofac. Surg.* **2021**, *122*, 241–247. [CrossRef] [PubMed]

34. Dragonas, P.; Prasad, H.S.; Yu, Q.; Mayer, E.T.; Fidel, P.L., Jr. Bone Regeneration in Maxillary Sinus Augmentation Using Advanced Platelet-Rich Fibrin (A-PRF) and Plasma Rich in Growth Factors (PRGF): A Pilot Randomized Controlled Trial. *Int. J. Periodontics Restor. Dent.* **2023**, *43*, 319–327. [CrossRef] [PubMed]
35. Ghonima, J.; El Rashidy, M.; Kotry, G.; Abdelrahman, H. The efficacy of combining advanced platelet-rich fibrin to biphasic alloplast in management of intrabony defects (randomized controlled clinical trial). *Alex. Dent. J.* **2020**, *45*, 8–13. [CrossRef]
36. Giudice, A.; Esposito, M.; Bennardo, F.; Brancaccio, Y.; Buti, J.; Fortunato, L. Dental extractions for patients on oral antiplatelet: A within-person randomised controlled trial comparing haemostatic plugs, advanced-platelet-rich fibrin (A-PRF+) plugs, leukocyte- and platelet-rich fibrin (L-PRF) plugs and suturing alone. *Int. J. Oral Implantol.* **2019**, *12*, 77–87.
37. Hartlev, J.; Erik Nørholt, S.; Spin-Neto, R.; Kraft, D.; Schou, S.; Isidor, F. Histology of augmented autogenous bone covered by a platelet-rich fibrin membrane or deproteinized bovine bone mineral and a collagen membrane: A pilot randomized controlled trial. *Clin. Oral Implant. Res.* **2020**, *31*, 694–704. [CrossRef] [PubMed]
38. Hartlev, J.; Nørholt, S.E.; Schou, S.; Isidor, F. Pain after mandibular ramus block harvesting and lateral ridge augmentation with and without involvement of platelet-rich fibrin: A randomized controlled trial. *Int. J. Oral Maxillofac. Surg.* **2021**, *50*, 384–390. [CrossRef] [PubMed]
39. Ivanova, V.; Chenchev, I.; Zlatev, S.; Kanazirski, N. Dimensional Ridge Alterations and Histomorphometric Analysis Following Socket Preservation with PRF or Allograft. Randomized Controlled Clinical Study. *J. IMAB–Annu. Proc. Sci. Pap.* **2019**, *25*, 2853–2861. [CrossRef]
40. Jayadevan, V.; Gehlot, P.M.; Manjunath, V.; Madhunapantula, S.V.; Lakshmikanth, J.S. A comparative evaluation of Advanced Platelet-Rich Fibrin (A-PRF) and Platelet-Rich Fibrin (PRF) as a Scaffold in Regenerative Endodontic Treatment of Traumatized Immature Non-vital permanent anterior teeth: A Prospective clinical study. *J. Clin. Exp. Dent.* **2021**, *13*, e463–e472. [CrossRef]
41. Liberati, A.; Altman, D.G.; Tetzlaff, J.; Mulrow, C.; Gøtzsche, P.C.; Ioannidis, J.P.A.; Clarke, M.; Devereaux, P.J.; Kleijnen, J.; Moher, D. The PRISMA statement for reporting systematic reviews and meta-analyses of studies that evaluate health care interventions: Explanation and elaboration. *Ann. Intern. Med.* **2009**, *151*, 65–94. [CrossRef] [PubMed]
42. Higgins, J.P.T.; Sterne, J.A.C.; Savović, J.; Page, M.J.; Hróbjartsson, A.; Boutron, I.; Reeves, B.; Eldridge, S. A revised tool for assessing risk of bias in randomized trials. In *Cochrane Methods: Cochrane Database of Systematic Reviews*; Chandler, J., McKenzie, J., Boutron, I., Welch, V., Eds.; John Wiley & Sons, Ltd.: Hoboken, NJ, USA, 2016; Volume 10, (Suppl. 1). [CrossRef]
43. Sterne, J.A.C.; Savović, J.; Page, M.J.; Elbers, R.G.; Blencowe, N.S.; Boutron, I.; Cates, C.J.; Cheng, H.Y.; Corbett, M.S.; Eldridge, S.M.; et al. RoB 2: A revised tool for assessing risk of bias in randomised trials. *BMJ* **2019**, *366*, l4898. [CrossRef] [PubMed]
44. Kalash, S.; Aboelsaad, N.; Shokry, M.; Choukroun, J. The efficiency of using advanced PRF-xenograft mixture around immediate implants in the esthetic zone: A randomized controlled clinical trial. *J. Osseointegr.* **2017**, *9*, 317–322. [CrossRef]
45. Lavagen, N.; Nokovitch, L.; Algrin, A.; Dakpe, S.; Testelin, S.; Devauchelle, B.; Gbaguidi, C. Efficiency of advanced-PRF usage in the treatment of alveolar cleft with iliac bone graft: A retrospective study. *J. Craniomaxillofac. Surg.* **2021**, *49*, 923–928. [CrossRef]
46. Nowak, J.M.; Surma, S.; Romańczyk, M.; Wojtowicz, A.; Filipiak, K.J.; Czerniuk, M.R. Assessment of the Effect of A-PRF Application during the Surgical Extraction of Third Molars on Healing and the Concentration of C-Reactive Protein. *Pharmaceutics* **2021**, *13*, 1471. [CrossRef] [PubMed]
47. Öngöz Dede, F.; Bozkurt Doğan, Ş.; Çelen, K.; Çelen, S.; Deveci, E.T.; Seyhan Cezairli, N. Comparison of the clinical efficacy of concentrated growth factor and advanced platelet-rich fibrin in the treatment of type I multiple gingival recessions: A controlled randomized clinical trial. *Clin. Oral Investig.* **2023**, *27*, 645–657. [CrossRef] [PubMed]
48. Pereira, D.A.; Mendes, P.G.J.; Prisinoto, N.R.; de Rezende Barbosa, G.L.; Soares, P.B.F.; de Oliveira, G.J.P.L. Advanced platelet-rich-fibrin (A-PRF +) has no additional effect on the healing of post-extraction sockets of upper third molars. A split mouth randomized clinical trial. *Oral Maxillofac. Surg.* **2023**, *27*, 411–419. [CrossRef]
49. Praganta, J.; De Silva, H.; De Silva, R.; Tong, D.C.; Thomson, W.M. Effect of Advanced Platelet-Rich Fibrin (A-PRF) on Postoperative Level of Pain and Swelling Following Third Molar Surgery. *J. Oral Maxillofac. Surg.* **2024**, *82*, 581–589. [CrossRef] [PubMed]
50. Rachna, M.; Nandita, S.; Rashmi, K.S.; Vidya, G.B.; Srikant, N.; Ananya, J.; Dharnappa, P. Eggshell membrane as a regenerative material in alveolar bone grafting in combination with advanced platelet rich fibrin. *Clin. Ter.* **2024**, *175*, 219–225. [CrossRef] [PubMed]
51. Şen, D.Ö.; Şengül, B.I.; Yarkaç, F.U.; Öncü, E. Impact of platelet-rich fibrin derivatives on patient morbidity and quality of life in palatal donor sites following free gingival graft surgery: A randomized clinical trial. *Clin. Oral Investig.* **2024**, *28*, 631. [CrossRef] [PubMed]
52. Soto-Peñaloza, D.; Peñarrocha-Diago, M.; Cervera-Ballester, J.; Peñarrocha-Diago, M.; Tarazona-Alvarez, B.; Peñarrocha-Oltra, D. Pain and quality of life after endodontic surgery with or without advanced platelet-rich fibrin membrane application: A randomized clinical trial. *Clin. Oral Investig.* **2020**, *24*, 1727–1738. [CrossRef]
53. Sousa, F.; Machado, V.; Botelho, J.; Proença, L.; Mendes, J.J.; Alves, R. Effect of A-PRF Application on Palatal Wound Healing after Free Gingival Graft Harvesting: A Prospective Randomized Study. *Eur. J. Dent.* **2020**, *14*, 63–69. [CrossRef]
54. Śmieszek-Wilczewska, J.; Balicz, A.; Morawiec, T.; Al-Maawi, S.; Heselich, A.; Sader, R.; Rutkowski, J.L.; Mourão, C.F.; Ghanaati, S. Effectiveness of Oroantral Communication Closure Using Solid Platelet-Rich Fibrin Compared to a Conventional Treatment Approach: A Randomized Clinical Trial. *J. Oral Implantol.* **2024**, *50*, 3–8. [CrossRef] [PubMed]

55. Tadepalli, A.; Chekurthi, S.; Kavassery Balasubramanian, S.; Parthasarathy, H.; Ponnaiyan, D. Comparative Evaluation of Clinical Efficacy of Leukocyte-Rich Platelet-Rich Fibrin with Advanced Platelet-Rich Fibrin in Management of Gingival Recession Defects: A Randomized Controlled Trial. *Med. Princ. Pract.* **2022**, *31*, 376–383. [CrossRef] [PubMed]
56. Torul, D.; Omezli, M.M.; Kahveci, K. Evaluation of the effects of concentrated growth factors or advanced platelet rich-fibrin on postoperative pain, edema, and trismus following lower third molar removal: A randomized controlled clinical trial. *J. Stomatol. Oral Maxillofac. Surg.* **2020**, *121*, 646–651. [CrossRef]
57. Trimmel, B.; Gyulai-Gaál, S.; Kivovics, M.; Jákob, N.P.; Hegedűs, C.; Szabó, B.T.; Dobó-Nagy, C.; Szabó, G. Evaluation of the Histomorphometric and Micromorphometric Performance of a Serum Albumin-Coated Bone Allograft Combined with A-PRF for Early and Conventional Healing Protocols after Maxillary Sinus Augmentation: A Randomized Clinical Trial. *Materials* **2021**, *14*, 1810. [CrossRef]
58. Yewale, M.; Bhat, S.; Kamath, A.; Tamrakar, A.; Patil, V.; Algal, A.S. Advanced platelet-rich fibrin plus and osseous bone graft for socket preservation and ridge augmentation—A randomized control clinical trial. *J. Oral Biol. Craniofac. Res.* **2021**, *11*, 225–233. [CrossRef]
59. Yüce, E.; Kömerik, N. Potential effects of advanced platelet rich fibrin as a wound-healing accelerator in the management of alveolar osteitis: A randomized clinical trial. *Niger. J. Clin. Pract.* **2019**, *22*, 1189–1195. [CrossRef] [PubMed]
60. Zahid, T.M.; Nadershah, M. Effect of Advanced Platelet-rich Fibrin on Wound Healing after Third Molar Extraction: A Split-mouth Randomized Double-blind Study. *J. Contemp. Dent. Pract.* **2019**, *20*, 1164–1170. [CrossRef]

Journal of
*Functional Biomaterials*

MDPI

*Article*

# Effect of Fluoride Varnishes on Demineralization and Acid Resistance in Subsurface Demineralized Lesion Models

**Rika Iwawaki [1], Taku Horie [1,2,†], Abdulaziz Alhotan [3], Yuka Nagatsuka [1], Keiko Sakuma [1], Kumiko Yoshihara [4,5] and Akimasa Tsujimoto [1,6,7,*,†]**

1 Department of Operative Dentistry, School of Dentistry, Aichi Gakuin University, Nagoya 464-8651, Japan; iwawaki@dpc.agu.ac.jp (R.I.); lifedoor@dpc.agu.ac.jp (T.H.); yuca@dpc.agu.ac.jp (Y.N.); virgo@dpc.agu.ac.jp (K.S.)
2 Department of Oral Health Sciences, BIOMAT, KU Leuven, 3000 Leuven, Belgium
3 Department of Dental Health, College of Applied Medical Sciences, King Saud University, P.O. Box 10219, Riyadh 12372, Saudi Arabia; aalhotan@ksu.edu.sa
4 Health and Medical Research Institute, National Institute of Advanced Industrial Science and Technology (AIST), Takamatsu 761-0395, Japan; kumiko.yoshihara@aist.go.jp
5 Department of Pathology & Experimental Medicine, Graduate School of Medicine, Dentistry and Pharmaceutical Sciences, Okayama University, Okayama 700-8558, Japan
6 Department of Operative Dentistry, University of Iowa College of Dentistry, Iowa City, IA 52242, USA
7 Department of General Dentistry, Creighton University School of Dentistry, Omaha, NE 68102, USA
* Correspondence: aki-tj@dpc.agu.ac.jp; Tel.: +81-52-751-7181
† These authors contributed equally to this work.

**Abstract:** This study aimed to clarify the effects of high-concentration fluoride varnish application on the inhibition of the progression of initial enamel caries. Remineralization capacity and acid resistance following high-concentration fluoride varnish application were compared with untreated models and models treated with fluoride mouthwash. Bovine enamel was used to create a model of initial enamel caries. The high-concentration fluoride varnishes Enamelast and Clinpro White Varnish and the fluoride mouthwash Miranol were used. Specimens were evaluated using Contact Microradiography (CMR) and an Electron Probe Micro-Analyzer (EPMA). While a single application of high-concentration fluoride varnish and short-term fluoride mouthwash use did not appear to cause remineralization in the subsurface demineralized layer, improvements in acid resistance were observed, leading to reduced demineralization under subsequent acidic challenges.

**Keywords:** dental biomaterials; surface analysis; enamel; fluoride; remineralization

## 1. Introduction

Dental caries are among the most prevalent infectious diseases in the world, affecting over 90% of adults according to a survey conducted by the US CDC from 2011 to 2016 [1]. This prevalence has not changed substantially since an earlier survey in 1999–2004. Caries are strongly influenced by the production of organic acids by cariogenic bacteria, and these acids demineralize the tooth structure. In initial enamel caries, subsurface demineralized lesions are formed while the superficial mineralized layers are retained, and these are clinically observed as white spots [2]. Generally, in the enamel surface, decalcification and recalcification occur repeatedly, and when this balance is disturbed and demineralization becomes predominant, the subsurface demineralized lesions progress and eventually develop into caries [3,4]. The concept of Minimal Intervention Dentistry (MID), originally proposed by the World Dental Federation (FDI) in 2002, and then modified in 2016, recommends the improvement of oral microflora through plaque control and sugar restriction, patient education through guidance on diet and oral hygiene, and remineralization therapy for such initial enamel caries [5]. The ICDAS (International Caries Detection & Assessment System), which was proposed in 2005, classifies caries in detail according to observable

features and classifies these initial enamel caries lesions as code I or II, recommending remineralization treatment and ongoing observation [6]. It is now broadly accepted that initial enamel caries can be treated through remineralization and that a surgical approach is not necessary except in cases of high esthetic requirements.

One approach to treating such initial enamel caries includes the use of fluoride varnish. These materials are easy to apply to the teeth and include colorings and flavorings that ensure that they are pleasant for the patient. They remain in place on the tooth for at least 24 h, even in thin layers, and are reported to continue releasing fluoride over that period. This material has been approved for treating dental hypersensitivity in the USA and Japan, and it is widely used "off-label" for caries prevention and to promote remineralization [7]. Fluoride varnishes have been used in Europe for sixty years, and the evidence for their safety is extremely strong [8]. Furthermore, clinical studies have shown that the application of these materials at routine check-ups can reduce the incidence of caries by between 25% and 45% in patients ranging from young children to adults [8]. In the UK, such treatment is available through the National Health Service up to twice a year, regardless of age. The effectiveness and safety of this approach are widely recognized in Europe, as shown by evidence from several Cochrane reviews [8]. This treatment is believed to promote both remineralization and acid resistance. However, although the epidemiological evidence for its effectiveness is very strong, understanding of its mechanism of action is still inadequate.

This study uses a laboratory initial enamel caries model to evaluate two possible mechanisms for the anticariogenic effect of this treatment: the remineralization ability of fluoride varnish, and the changes in acid resistance of the area to which it is applied.

## 2. Materials and Methods

### 2.1. Preparation of Enamel Specimens

Three specimens per tooth were cut from the labial surface of freshly extracted bovine anterior teeth (bought from Kenis Limited (Osaka, Japan), https://global.kenis.co.jp) using a low-speed cutting machine (Isomet, Buehler, Lake Bluff, IL, USA), which were 4 mm wide in the direction of the tooth axis, 5 mm long perpendicular to the tooth axis, and 4 mm thick. The specimens were then polished to #2000 under water using silicon carbide paper so that the labial enamel surfaces were as flat as possible. The pulpal surface of the specimen was then removed parallel to the enamel surface so that the final thickness of the specimen was 3 mm. The specimens were then ultrasonically cleaned in distilled water for 1 min and masked with nail varnish to create a 3 × 4 mm rectangular surface in the center of the labial enamel, which was used as the treatment surface (Figure 1).

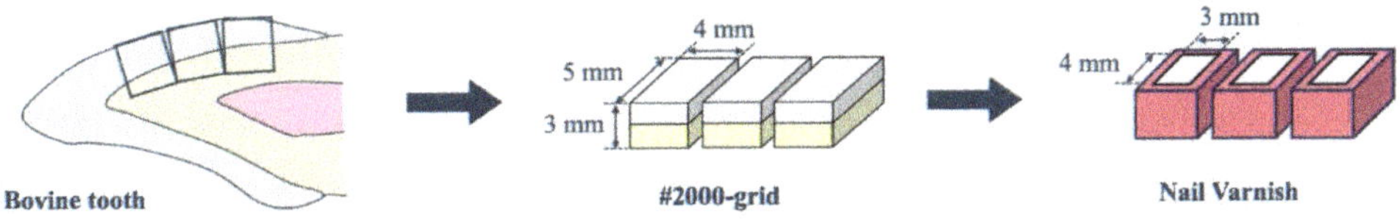


**Figure 1.** The preparation of enamel specimens.

### 2.2. Subsurface Demineralization Model

Specimens with a subsurface demineralized layer were prepared for different remineralization treatments. The preparation for making specimens with a subsurface demineralized layer was carried out using lactic acid following the methods of Hayashi [9]. The surface was immersed in 20 mL (37 °C) of 8% Methocel MC gel (Fluka, Buchs, Switzerland) for 24 h, with the treated surface facing upwards, and then a sheet of filter paper of approximately the same size as the inner diameter of the container was laid over the surface of the gel and 20 mL of 0.1 mol/L lactic acid solution (pH 4.6) was gently applied to the paper. The specimens were demineralized for 10 days to make the subsurface demineralized model (Figure 2).

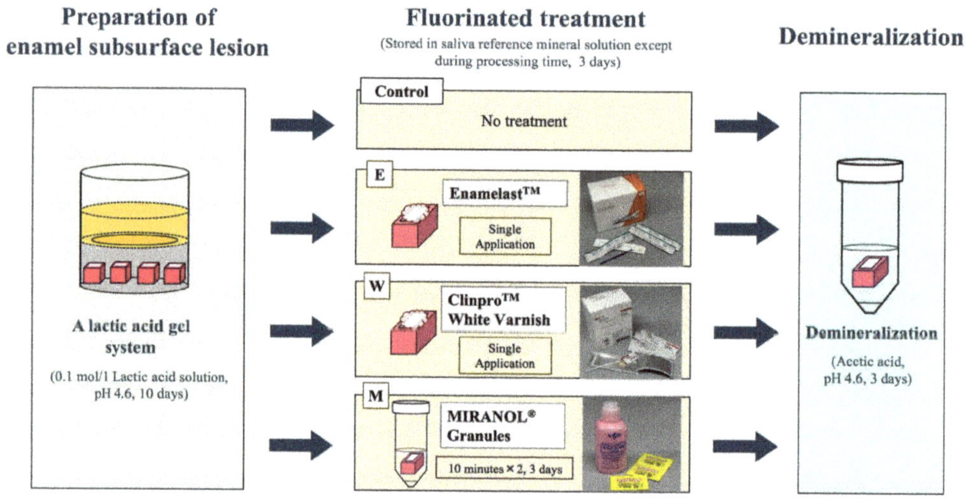


**Figure 2.** The procedures for preparation of enamel subsurface lesion, fluorinated treatment, and demineralization.

### 2.3. Fluoride Treatment

The subsurface demineralized models were randomly divided into four groups. The first group (E group) was treated with Enamelast (Ultradent Products, South Jordan, UT, USA) according to the manufacturer's instructions. The second group (W group) was treated with Clinpro White Varnish (Solventum, St. Paul, MN, USA), again according to the manufacturer's instructions. The third group (M group) was immersed in Miranol (Bee Brand Medico Dental, Osaka, Japan) for ten minutes twice a day for a period of three days. The final group was left untreated as a control group. All specimens were stored in artificial saliva (Saliveht Aerosol, Teijin Pharma, Tokyo, Japan) for 3 days, except when they were being treated. The surfaces of these specimens were finally cleaned with distilled water.

### 2.4. Re-Demineralization

The fluoride-treated samples were immersed into an artificial demineralizing solution (50 mmol/L acetic acid, 1.5 mmol/L $CaCl_2$, 0.9 mmol/L $KH_2PO_4$, pH 4.6, 37 °C) for 3 days [9].

### 2.5. CMR and EPMA Analysis

All prepared samples of each group of specimens were dehydrated with an ethanol system and embedded in epoxy resin (Polysciences, Warrington, PA, USA). A thin section of each specimen was prepared for Contact Microradiogram (CMR) analysis through extraction of the central part of the specimen parallel to the tooth axis using a low-speed cutting machine. The bulk sections remaining after the extraction of the thin section were used for EPMA analysis. The sections extracted for CMR were prepared with a thickness of approximately 300 µm, and they were refined into 100 µm thick sections by finishing with #2000 SiC paper. The sectioned surfaces of the other remaining blocks were also polished up to #2000 SiC paper for Electron Probe Micro-Analyzer (EPMA) analysis (Figure 3).

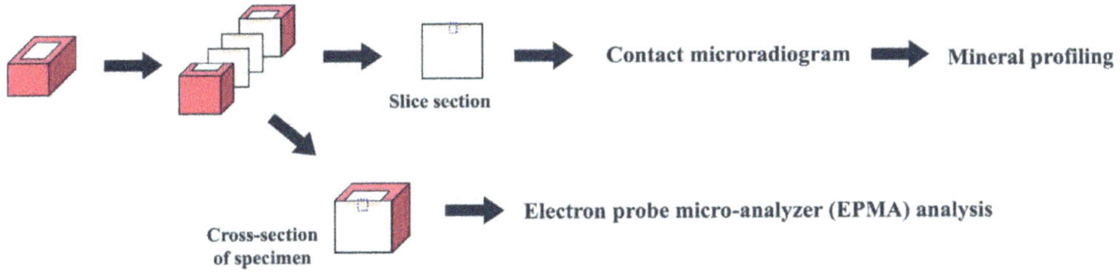


**Figure 3.** The observation and analysis.

### 2.6. CMR Imaging

Polished sections of each group were subjected to CMR photography using a soft X-ray non-destructive inspection system with an acceleration voltage of 10 kV, a tube current of 2 mA, and an irradiation time of 5 min, avoiding drying of the sections as much as possible, together with aluminum step wedges for calibration curves. A band at the center of the section approximately 100 μm wide was imaged over a distance of 250 μm from the enamel surface edge of the specimen, representing a depth of 250 μm from the surface of the treated sample. A glass dry plate (High Precision Photo Plate, Konica Minolta, Tokyo, Japan) was used for imaging, and a special developing solution (CDH-100, Konica Minolta) and a fixing solution (CFL-881, Konica Minolta) were used for developing and fixing, respectively.

### 2.7. Mineral Profiling

The CMR of each group was digitized using a camera (DP70, Olympus, Tokyo, Japan) connected to an optical microscope (SZX9, Olympus), and a mineral profile was created using image analysis software (WinROOF Ver3.4,0, Mitani Corporation, Fukui, Japan). The gray values were measured in 256 shades of gray. The obtained gray values were corrected with reference to the gray value of the corresponding aluminum step wedge.

The gray value of the area on the film where the tooth sample was not present was taken to represent 0 vol% of mineral, and the gray value of the area where sound enamel was visible was taken to represent 100 vol% of mineral. These values were used to calculate the demineralization at each depth of the sample, and the amount of mineral loss ($\Delta Z$, vol%·μm) was calculated. In other words, the amount of mineral loss was calculated from the area between the graph for one condition and another. The mineral losses ($\Delta Z$) between fluoride and re-demineralization models were measured.

### 2.8. EPMA Observation

The specimens for EPMA analysis of each group were mirror polished with 0.3 μm alumina oxide and ultrasonically cleaned in distilled water for 1 min. The surface was then treated with carbon deposition, and the microstructure of the longitudinal section was observed using a field emission electron probe micro-analyzer (FE-EPMA, JXA 8530F, Jeol, Tokyo, Japan) at an acceleration voltage of 10 kV. The wavelength range of detected X-rays was 0.087 to 9.3 nm, with a 3 nm resolution for the secondary electron imaging.

### 2.9. Elemental Analysis

In order to investigate the distribution of minerals on the surface where the microstructure was observed, a surface analysis (acceleration voltage 10 kV, acceleration current 3.0 × 1r8A) was carried out for Ca and P using FE-EPMA.

## 3. Results

### CMR and Mineral Profiles

Figures 4 and 5 show the typical CMR and mineral profiles for each of the various fluoride treatment models.

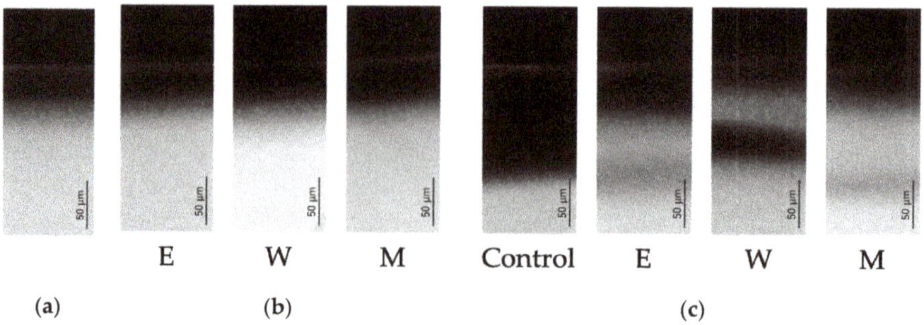


**Figure 4.** The typical CMRs. (**a**) Prepared enamel subsurface lesion. (**b**) After fluorinated treatment. (**c**) After demineralization.

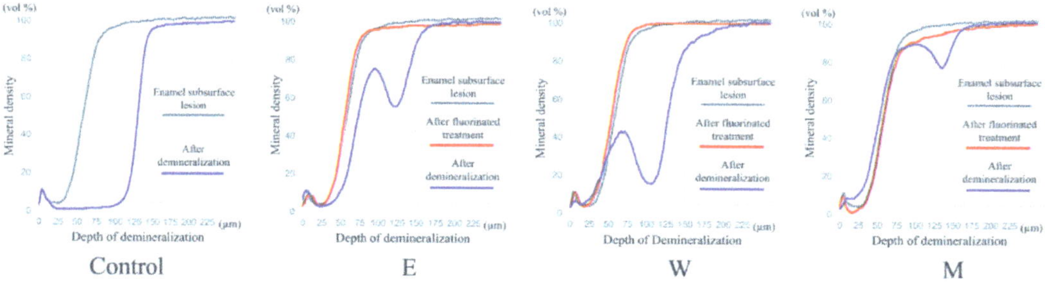


**Figure 5.** Typical mineral profiles.

Figure 6 shows the typical EPMA analysis for each of the various fluoride treatment models.

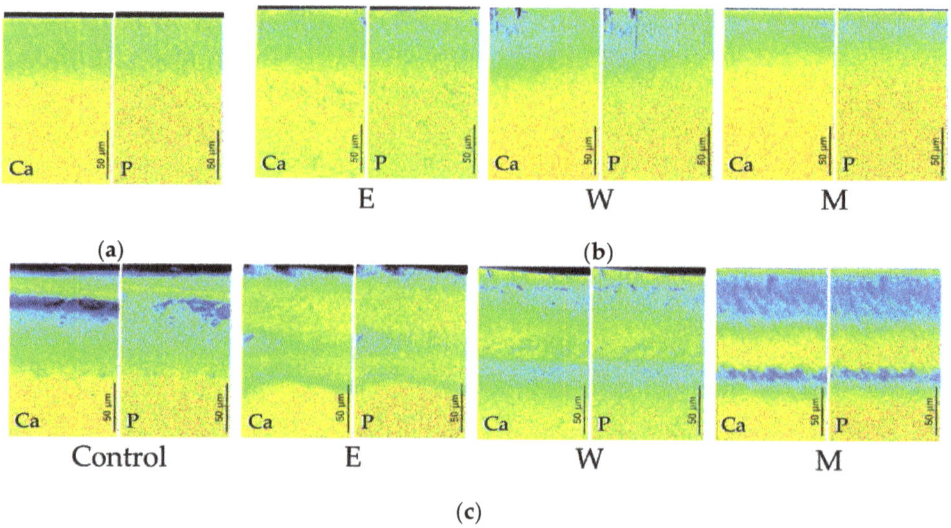


**Figure 6.** Typical EPMA analysis. (**a**) Prepared enamel subsurface lesion. (**b**) After fluorinated treatment. (**c**) After demineralization.

In the model of enamel subsurface demineralization, the typical profile of such demineralization was observed, with mineral loss just below the outermost layer of enamel,

although the outermost layer of enamel was retained. The mineral density was lowest around a depth of about 30 μm in the subsurface demineralized layer, and the mineral density at that depth was 4.0 vol%.

Using these mineral profiles as a reference, we defined the area from the outermost layer to a mineral density of 50% in the demineralized layer model, which extended from the outermost layer to a depth of around 60 μm, as the "sub-surface demineralised layer", and the area from a density of 50% to the intact part, which extended from a depth of around 60 μm to around 100 μm, as the "deep sub-surface demineralised layer".

In the fluoride application model, the CMR and mineral profiles of the M, W, and E groups were the same as those of the enamel subsurface demineralization model.

On the other hand, in the re-demineralization model, different CMR and mineral profiles were observed depending on the type of fluoride treatment. In the control group, although the outermost layer of enamel was retained, the CMR of the deep subsurface demineralized layer was observed to be significantly demineralized, and mineral loss was observed in the mineral profile to a depth of 50 μm below this layer.

In groups M, W, and E, the outermost layer of enamel and the deep subsurface demineralized layer were preserved in the CMR, but transparency was observed in the lower part of the subsurface demineralized layer. On the other hand, the trends in the mineral profiles of groups M, W, and E were different.

In the M group, the mineral density of the subsurface demineralized layer was maintained, but a decrease in mineral density was observed in the layer below it, with the greatest loss of mineral density occurring at a depth of around 130 μm, where the mineral density was 77.3 vol%.

In the mineral profiles of groups W and E, a decrease in mineral density in the subsurface demineralized layer was observed, but the amount of mineral loss was smaller than in the control group. On the other hand, a marked decrease in mineral density was observed in the lower layer, with the greatest mineral density loss occurring at a depth of around 100 μm in the W group and around 120 μm in the E group, with mineral densities of 15.5 vol% and 55.9 vol%, respectively.

The mineral loss (vol %·μm) for the control, E, W, and M groups in these representative cases was 7410.0, 3353.5, 5589.5, and 554.3, respectively.

In the EPMA analysis, similarly to mineral changes observed in CMR, loss of Ca and P was identified in the subsurface enamel layer. The distributions of Ca and P were almost completely identical.

## 4. Discussion

It has been widely reported that the application of fluoride to demineralized lesions in the enamel surface layer promotes remineralization [10]. However, in the present experiment, there was no recovery of mineralization or increase in Ca and P concentrations in the subsurface demineralized layer in any of the E, W, or M groups in the fluoride application model. Previous studies have reported that for fluoride to promote remineralization, a continuous supply of calcium and fluoride ions contained in saliva is necessary. Therefore, it was thought unlikely that remineralization of the subsurface demineralized layer would be promoted by fluoride application over the short period of time (3 days) covered in this study. In addition, when fluoride is applied to the subsurface demineralized layer of the enamel, the fluoride ions may replace hydroxide ions in the enamel and remain in the superficial layer. Therefore, although immersion in artificial saliva (mineral solution) for a long period of time may promote remineralization through reactions with the calcium and phosphate ions contained in it, and there is a possibility that remineralization will be observed under these conditions, further investigation is needed.

The condition of the demineralized enamel samples that were treated with fluoride and then re-demineralized showed different trends in each of the E, W, and M groups compared to the control group. It is generally known that the application of fluoride improves resistance to acid in healthy enamel as well as in the subsurface demineralized layer. In this study, too, the M, W, and E groups formed a layer that showed resistance to acid in the lower part of the subsurface demineralized layer, and the deeper part of the sample was selectively demineralized, unlike the control groups. Hayashi et al. reported that when the subsurface demineralized layer of enamel that had undergone remineralization treatment was demineralized again, the surface layer and the subsurface demineralized layer acquired acid resistance, and the deeper layer below was more strongly attacked by the acid [9]. The results of this study are consistent with previous reports, which argue that fluorapatite, which is resistant to acid, is formed through the replacement of hydroxide ions in demineralized hydroxyapatite with fluoride ions [11]. The reason for demineralization directly beneath the layer with acid resistance was thought to be that demineralization starts in the deeper layers of enamel that were not reached by the fluoride, and so that part is selectively demineralized.

Turning to CMR imaging, there was no clear difference between the images of the subsurface demineralized samples from the control group and from the treatment groups. Fluorine (F) is generally considered to have low X-ray contrast, so even if fluoride compounds accumulated on the surface of the demineralized area after fluoride application, it is likely difficult to observe this using CMR. On the other hand, when the demineralized layer was re-demineralized, the elution of calcium ions was suppressed in the layer where fluorapatite was formed, and it was expected that this would cause a difference in the amount of minerals measured using CMR, as was observed.

When comparing the effects of different fluoride treatments on inhibiting demineralization, the results of this study showed that the mineral loss in groups W (Clinpro Varnish) and E (Enamelast) was greater than in group M (Miranol), despite the fact that the fluoride concentration of the fluoride varnishes, Clinpro Varnish and Enamelast, was higher than that of the fluoride mouthwash, Miranol, at 22,600 ppm. Miranol, a mouthwash, is a solution containing a low concentration of fluoride (250 ppm F), and it can be thought that the fact that it was more easily able to penetrate into the demineralized area under the enamel surface made it more effective than the high-concentration, but high-viscosity, fluoride varnishes in suppressing mineral loss. According to a previous report on the amount of fluoride ions released from fluoride varnish, the amount of fluoride released when Clinpro Varnish was immersed in water for 24 h was $74.0 \pm 32.2$ μmol/g [12]. This suggests that only about 1/16 of the fluoride, which is present at a high concentration of 22,600 ppm in the material itself, was actually released. Additionally, as this fluoride is diluted in the surrounding water, the concentration of ions available to react with the demineralized layer is likely to be even lower. On the other hand, although the fluoride concentration in Miranol is low at 250 ppm, it is all in aqueous solution, suggesting that the ion concentration available to react with the demineralized layer may be higher in the mouthwash than in the varnish.

When fluoride treatment is applied to the subsurface demineralized layer, the acquisition of an acid-resistant layer and remineralization are also influenced by contact with the mineral components contained in saliva, so, in this experiment, the specimens were immersed in artificial saliva in order to more accurately reproduce the intraoral environment. In other words, the mineral components contained in saliva are essential for the acquisition of an acid-resistant layer and remineralization of the enamel through a chemical reaction with fluoride ions [13]. Bolis et al. reported that the amount of fluoride uptake by enamel from a high-concentration fluoride varnish applied directly to it does not necessarily correlate with the amount of fluoride released in solution [14]. In addition, Kim et al. reported that the improvement in surface hardness was greater when fluoride varnish was applied indirectly to the enamel compared to when it was applied directly, suggesting that fluoride varnish may prevent calcium and phosphate ions in saliva from penetrating into demineralized areas by covering the enamel [15]. In other words, fluoride varnish includes substances, such as rosin, that secure both adhesiveness and insolubility, and although it can remain in the area where it is applied for a long time, it may be that the increase in fluoride concentration in the surrounding area is limited and that the layer of varnish may also reduce the supply of other ions to that area. Godoi et al. studied the effect of fluoride varnishes on remineralization, using hardness as a proxy for mineral content, and they also found no remineralization of the subsurface lesion [16].

In this study, we conducted experiments in the hope that a single application of high-concentration fluoride varnish would result in high acid resistance and remineralization, but no significant remineralization was observed in either group, and the creation of acid-resistant layers was less effective than that with fluoride-containing mouthwash. However, it is worth noting that a significant improvement in acid resistance was observed in the group that received a single short application of fluoride varnish compared to the untreated group. Because fluoride mouthwashes require continuous application by the patients themselves, it is possible that a single application of fluoride varnish may be an effective way to secure some increase in acid resistance in patients who may not cooperate with using a mouthwash, such as young children, or who have difficulty with rinsing, such as the elderly.

## 5. Conclusions

The application of high-concentration fluoride varnish or fluoride mouthwash to an early enamel caries model was found to not significantly promote remineralization in the subsurface demineralized layer, but it improved the acid resistance of the deepest part of the layer.

**Author Contributions:** Conceptualization, T.H. and A.T.; methodology, T.H. and Y.N.; software, T.H. and Y.N.; validation, R.I., T.H., Y.N., K.S. and K.Y.; formal analysis, T.H. and K.Y.; investigation, T.H.; resources, A.T.; data curation, T.H.; writing—original draft preparation, T.H. and A.T.; writing—review and editing, T.H., A.A. and A.T.; visualization, R.I. and T.H.; supervision, A.T.; project administration, A.T.; funding acquisition, T.H., A.A. and A.T. All authors have read and agreed to the published version of the manuscript.

**Funding:** This research was supported by a grant from Ultradent Japan (R05-32, 2023–2024) and the Researchers Supporting Project number (RSPD2024R790), King Saud University, Riyadh, Saudi Arabia.

**Institutional Review Board Statement:** Not applicable.

**Informed Consent Statement:** Not applicable.

**Data Availability Statement:** The original contributions presented in the study are included in the article, further inquiries can be directed to the corresponding author.

**Acknowledgments:** The authors thank the grant from Ultradent Japan (R05-32, 2023–2024) and the Researchers Supporting Project number (RSPD2024R790), King Saud University, Riyadh, Saudi Arabia.

**Conflicts of Interest:** The funding sponsors had no role in the design of the study; in the collection, analyses, or interpretation of data; in the writing of the manuscript, and in the decision to publish the results. The authors declare no conflict of interest.

## References

1. Centers for Disease Control and Prevention. *Oral Health Surveillance Report: Trends in Dental Caries and Sealants, Tooth Retention, and Edentulism, United States, 1999–2004 to 2011–2016*; Centers for Disease Control and Prevention, U.S. Department of Health and Human Services: Washington, DC, USA, 2019.
2. Ten Cate, J.M.; Duijsters, P.P. Influence of fluoride in solution on tooth demineralization. *Caries Res.* **1983**, *17*, 193–199. [CrossRef] [PubMed]
3. Featherstone, J.D. The science and practice of caries prevention. *J. Am. Dent. Assoc.* **2000**, *131*, 887–899. [CrossRef] [PubMed]
4. Ingram, G.S.; Silverstone, L.M. A chemical and histological study of artificial caries in human dental enamel in vitro. *Caries Res.* **1981**, *15*, 393–398. [CrossRef] [PubMed]
5. FDI Policy Statement. Minimal Intervention Dentistry (MID) for Managing Dental Caries. Available online: https://www.fdiworlddental.org/minimal-intervention-dentistry-mid-managing-dental-caries (accessed on 10 November 2024).
6. Ismail, A.I.; Sohn, W.; Tellez, M.; Amaya, A.; Sen, A.; Hasson, H.; Pitts, N.B. The International Caries Detection and Assessment System (ICDAS): An integrated system for measuring dental caries. *Community Dent. Oral Epidemiol.* **2007**, *35*, 170–178. [CrossRef] [PubMed]

7. Garcia, R.I.; Gregorich, S.E.; Ramos-Gomez, F.; Braun, P.A.; Wilson, A.; Albino, J.; Tiwari, T.; Harper, M.; Batliner, T.S.; Rasmussen, M.; et al. Absence of Fluoride Varnish–Related Adverse Events in Caries Prevention Trials in Young Children, United States. *Prev. Chronic Dis.* **2017**, *14*, 160372. [CrossRef] [PubMed]
8. Marinho, V.C.; Worthington, H.V.; Walsh, T.; Clarkson, J.E. Fluoride varnishes for preventing dental caries in children and adolescents. *Cochrane Database Syst. Rev.* **2013**, *7*, CD002279. [CrossRef] [PubMed]
9. Hayashi, M. Characteristic changes of enamel surface layer following demineralization of remineralized enamel subsurface lesion. *Jpn. J. Conserv. Dent.* **2012**, *55*, 398–410.
10. Whelton, H.P.; Spencer, A.J.; Do, L.G.; Rugg-Gunn, A.J. Fluoride revolution and dental caries: Evolution of policies for global use. *J. Dent. Res.* **2019**, *98*, 837–846. [CrossRef] [PubMed]
11. de Leeuw, N.H. Resisting the onset of hydroxyapatite dissolution through the incorporation of fluoride. *J. Phys. Chem. B* **2004**, *108*, 1809–1811. [CrossRef]
12. Cochrane, N.J.; Shen, P.; Yuan, Y.; Reynolds, E.C. Ion release from calcium and fluoride containing dental varnishes. *Aust. Dent. J.* **2014**, *59*, 100–105. [CrossRef] [PubMed]
13. Farooq, I.; Bugshan, A. The role of salivary contents and modern technologies in the remineralization of dental enamel: A narrative review. *F1000Research* **2020**, *9*, 171. [CrossRef] [PubMed]
14. Bolis, C.; Härtli, G.P.; Lendenmann, U. Fluoride varnishes—Is there a correlation between fluoride release and deposition on enamel? *Oral Health Prev. Dent.* **2015**, *13*, 545–556. [PubMed]
15. Kim, H.N.; Kim, J.B.; Jeong, S.H. Remineralization effects when using different methods to apply fluoride varnish in vitro. *J Dent. Sci.* **2018**, *13*, 360–366. [CrossRef] [PubMed]
16. Godoi, F.A.; Carlos, N.R.; Bridi, E.C.; Amaral, F.L.B.D.; França, F.M.G.; Turssi, C.P.; Kantovitz, K.R.; Basting, R.T. Remineralizing effect of commercial fluoride varnishes on artificial enamel lesions. *Braz. Oral Res.* **2019**, *33*, e044. [CrossRef] [PubMed]

Journal of
*Functional Biomaterials*

*Article*

# Wear Resistance of Light-Cure Resin Luting Cements for Ceramic Veneers

**Miyuki Oshika [1,†], Takafumi Kishimoto [1,†], Taku Horie [1,2], Abdulaziz Alhotan [3], Masao Irie [4], Veronica C. Sule [5], Wayne W. Barkmeier [5] and Akimasa Tsujimoto [1,5,6,*]**

1 Department of Operative Dentistry, School of Dentistry, Aichi Gakuin University, Nagoya 464-8651, Japan; oshika@dpc.agu.ac.jp (M.O.); taka-ki@dpc.agu.ac.jp (T.K.); lifedoor@dpc.agu.ac.jp (T.H.)
2 Department of Oral Health Sciences, BIOMAT, KU Leuven, 3000 Leuven, Belgium
3 Department of Dental Health, College of Applied Medical Sciences, King Saud University, P.O. Box 10219, Riyadh 12372, Saudi Arabia; aalhotan@ksu.edu.sa
4 Department of Biomaterials, Okayama University Graduate School of Medicine, Dentistry and Pharmaceutical Sciences, 2-5-1 Shikata-cho, Kita-ku, Okayama 700-8525, Japan; mirie@md.okayama-u.ac.jp
5 Department of General Dentistry, Creighton University School of Dentistry, Omaha, NE 68102, USA; veronicasule@creighton.edu (V.C.S.); waynebarkmeier@creighton.edu (W.W.B.)
6 Department of Operative Dentistry, University of Iowa College of Dentistry, Iowa City, IA 52242, USA
* Correspondence: aki-tj@dpc.agu.ac.jp; Tel.: +81-52-751-7181
† These authors contributed equally to this work.

**Abstract:** The purpose of this study was to compare the wear resistance of light-cure resin luting cements for veneers with that of other luting materials investigated in earlier studies. An Alabama wear-testing machine was used to measure the wear resistance of four recent light-cure resin luting cements for veneers (G-Cem Veneer; Panavia V5 LC; RelyX Veneer Cement; and Vario-link Esthetic LC). The volume loss ranged from $0.027 \pm 0.003$ to $0.119 \pm 0.030$ mm$^3$, the mean facet depth from $56.053 \pm 7.074$ to $81.531 \pm 7.712$ μm, and the maximum facet depth from $100.439 \pm 26.534$ to $215.958 \pm 27.320$ μm. G-Cem Veneer showed significantly better ($p < 0.05$) wear resistance than the other materials tested. Representative SEM images were obtained which showed differences in form among the wear facets for the luting cements examined. Correlations were calculated between the three measurements for each material, and the pattern of correlations was also different for each material.

**Keywords:** dental biomaterials; surface analysis; wear; cementation; veneer

## 1. Introduction

Ceramic veneers are indirect aesthetic restorations with minimal tooth reduction, used to enhance the color, form, and function of unaesthetic teeth [1]. Light-cure resin luting cements are used for the bonding of veneers because the composition of the cements has better color stability than dual-cure resin luting cements [2]. These light-cure resin luting cements are used with primers, and therefore acidic functional monomers are not included in the cement itself. This is thought to contribute to the chemical stability of the material over the long term, and to play an important role in the quality of the bond between the veneer and the underlying tooth substrate. There is a long history and there have been long-term clinical studies on the use of ceramic veneers, and those studies showed excellent clinical performance of ceramic veneers over periods of 20 years, with more than 80% clinical survival rates [3,4].

However, one of the problems of long-term usage of veneers (3 surfaces) is gap formation between the veneer and the surrounding tooth structure, as shown in Figure 1, which shows a veneer after 18 years of clinical service. Although the veneer is still sound and shows good adaptation, the margin on the lingual side shows a clear gap between the veneer and the tooth substrate. This gap formation is mainly due to wear of the light-cure resin luting cement. This sort of gap is an ideal site for the retention of plaque, is vulnerable to staining, and may serve as a starting point of failure [5]. Also, if the worn surface of the cement becomes rougher, the wear rate will be accelerated by the abrasiveness of the surface [6]. These problems reduce the quality of the restoration even if they do not lead to failure, and therefore it is preferable to use a cement that minimizes them.

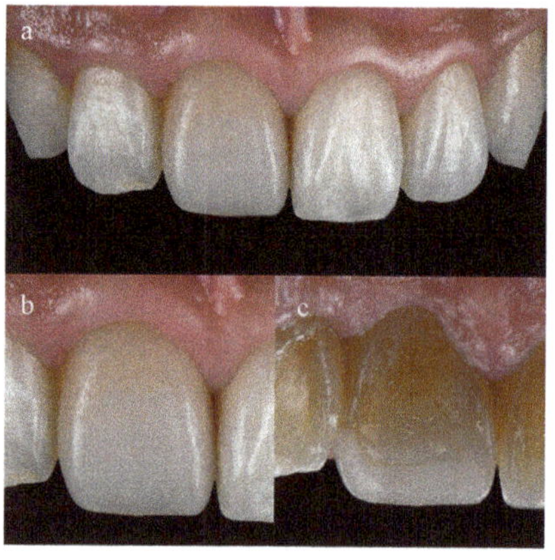


**Figure 1.** Lithium disilicate press ceramic veneer for tooth #8 after 18 years of usage: (**a**): frontal view; (**b**) enlarged facial view for tooth #8; and (**c**) enlarged lateral view with gap formation in cement line between the veneer and surrounding tooth structure. (Patient is author A.T.).

Recent developments in ceramics have made this problem even more pressing. Stronger ceramics, such as leucite and lithium disilicate reinforced ceramics, and translucent zirconia, have enabled the use of ultra-thin, thin and partial ceramic veneers with minimal preparations [7], and the creation of occlusal veneers, which do not extend all the way down the tooth [8]. With conventional crowns, the margin of the crown was mainly located in the gingival or subgingival area, and thus did not create much of an aesthetic or wear problem [9]. However, with anterior (three surfaces) or occlusal veneers, the cement line is visible on the outer surface of the tooth, and thus wear at that location may create aesthetic problems in addition to clinical issues. Furthermore, the joint line may be more exposed to wear during mastication, increasing the degree of wear beyond that seen with conventional crowns. Wear of the cement line on such restorations may also create an irregularity on the exposed surface of the tooth, which could in turn accelerate the wear of the antagonist surface.

This has made clinicians more concerned about the wear resistance of resin luting cements used to secure the ceramic veneer to tooth structures. Generally, the use of light-cure resin luting cements has been recommended for partial coverage restorations in order to avoid this kind of problem [2]. Although dual-cure resin luting cements are similar in many ways to light-cure versions, they are vulnerable to discoloration over the medium to long term, as most of the materials contain amines for the chemical activation of polymerization. Therefore, they were not preferred options for use in partial coverage aesthetic restorations [10]. The wear resistance of restorative resin-based composites has improved substantially in recent years [11]. Flowable resin-based composites (RBCs) have also been used [12], but similarly face several problems. First, their viscosity is still often higher than that of cement, particularly for injectable RBCs, which makes it difficult to obtain good adaptations and fill-thickness. Second, it has been generally considered that their wear resistance is still inferior to that of paste-type RBCs [13]. Third, the use of flowable RBCs for ceramic restorations generally provides less reliable bonding than that of a dual-cure resin luting cement [14]. Some clinicians have therefore tried using paste-type RBCs, but in this case it is necessary to heat the material to around 54 °C in order to reduce the viscosity to a point where it can be applied [15]. This heating has unknown effects on the properties of the RBC itself, and is also high enough to potentially have adverse effects on the pulp through the thin layer of enamel and dentin remaining. Furthermore, RBCs are not cements, and thus not ideally suited to this role.

These problems could be avoided if there were light-cured resin luting cements with similar wear resistance to the other classes of material. In the past, the Alabama wear-testing machine has been used to measure the wear resistance of materials in all these classes [16–18]. Using the same methodology ensures the comparability of results for different materials, but there are no reported studies using this technique to investigate the wear resistance of recently introduced light-cured resin luting cements.

Therefore, the purpose of this study is to evaluate the wear resistance of recently introduced light-cure resin luting cements through measurements of volume loss and both mean and maximum depth after simulated wear in an Alabama wear-testing machine, and compare with the alternative materials to determine whether light-cure resin luting cements are a reasonable clinical choice.

## 2. Materials and Methods

### 2.1. Study Materials

Four veneer cements: (1) G-Cem Veneer (GV, GC, Tokyo, Japan); (2) Panavia V5 LC (PV, Kuraray Noritake Dental, Tokyo, Japan); (3) RelyX Veneer Cement (RV, 3M ESPE, Seefeld, Germany); and (4) Variolink Esthetic LC (VE, Ivocla Vivadent, Schaan, Liechtenstein), were used in this laboratory investigation (Table 1).

**Table 1.** Composition of tested Light-cure Resin Luting Cements.

| Materials | Main Components | Manufacture | Study Code |
|---|---|---|---|
| G-Cem Veneer | UDMA, Esterification products of 4,4′-isopropylidenediphenol, ethoxylated and 2-methylprop-2-enoic acid, (octahydro-4,7-methano-1H-indenediyl)bis(methylene) bismethacrylate, 2,2-dimethyl-1,3-propanediyl bismethacrylate, 1,3,5-Triazine-2,4,6-triamine, polymer with formaldehyde, diphenyl(2,4,6-trimethylbenzoyl)phosphine oxide, 2,2′-ethylenedioxydiethyl dimethacrylate, Butylated hydroxytoluene, 2-(2H-benzotriazol-2-yl)-p-cresol, 6-tert-butyl-2,4-xylenol | GC | GV |
| Panavia Veneer LC | Silanated spherical silica filler, UDMA, Ytterbium trifluorideTriethyleneglycol dimethacrylate, Hydrophilic aliphatic dimethacrylate, Hydrophilic amide monomer, Accelerators, dl-Camphorquinone, Pigments | Kuraray Noritake | PV |
| RelyX Veneer Cement | Silane Treated Ceramic, BISGMA, TEGDMA, Silane Treated Silica, Reacted Polycaprolactone Polymer, Titanium Dioxide Diphenyliodonium Hexafluorophosphate, N,N-DIMETHYLBENZOCAINE, DL-Camphorquinone | 3M ESPE | RV |
| Variolink Esthetic LC | Matrix: urethane dimethacrylate and further methacrylate monomers.Inorganic fillers: ytterbium trifluoride and spheroid mixed oxide.Additional: initiators, stabilizers and pigments. | Ivolar Vivadent | VE |

### 2.2. Specimen Preparation

Previous research using an Alabama wear testing machine (University of Alabama at Birmingham, Birmingham, AL, USA) used a range of sample sizes, from 6 to 20 specimens in each group [16–18] (Figure 2). The sample size was determined based on effect size = 0.25 (medium) or 0.5 (large), a = 0.05, power = 0.8, and number of groups = 4. The result was that 180 specimens (medium effect size) or 48 (large effect size) were needed. With four groups, either 12 or 45 specimens were needed in each group, and 12 was selected.

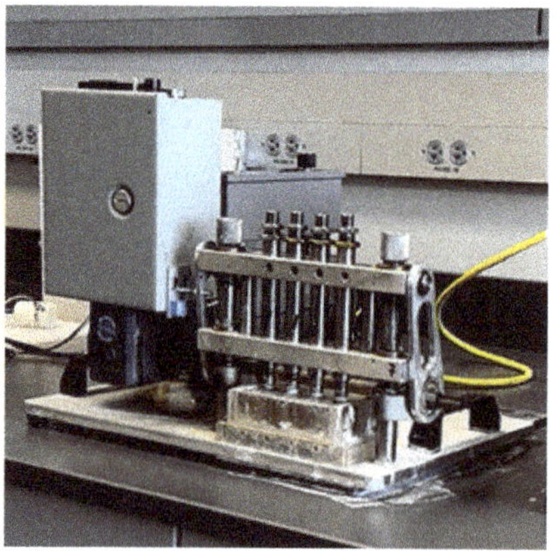

**Figure 2.** Alabama wear-testing machine in Creighton University School of Dentistry.

The samples were prepared following the methods described by Tsujimoto et al. [16]. Polishing, as described in reference [16], was conducted in order to approximate the clinical situation, in which these materials are polished.

### 2.3. Wear Testing

An Alabama wear-testing machine was used in this study [16–18], and the protocol described by Tsujimoto et al. [16] was followed.

### 2.4. Measurement of Wear Facet

The wear facets were measured as described in Tsujimoto et al. [16].

### 2.5. Statistical Analysis

A commercial statistical software package (IBM SPSS version 29.0.2.0, Chicago, IL, USA) was used. The Shapiro–Wilk test was used to confirm that the data were normally distributed. A one-way ANOVA (analysis of variance) with Tukey's post hoc honest significant difference test was used. VL MD and AD were analyzed with a significance level of 0.05. Additionally, Pearson correlation analysis was performed for all materials as well as between the VL, MD, and AD for each material.

### 2.6. Scanning Election Microscopy (SEM) Observations

Samples were chosen randomly for SEM observation using a Tabletop Microscopes (TM3000, Hitachi, Tokyo, Japan). A thin coating of an 80/20 gold–palladium alloy was applied in a sputter coater (Emitech SC7620 Mini Sputter Coaterk, Quorum Technologies, East Sussex, UK) to prevent accumulation of electrostatic charge at the sample surface. An operating voltage of 15 kV and magnifications of 100× and 2500× were used.

## 3. Results

### 3.1. Localized Wear

The results for localized wear (VL, MD and AD (Mean (average) depth)) are presented in Table 2. The one-way ANOVA among VL, MD and AD showed a significant difference ($p < 0.05$) for the material factor. Tukey's post hoc test for VL, MD and AD showed significant differences ($p < 0.05$) in localized wear among the materials tested. The VLs of the materials evaluated in this study ranged from 0.027 ± 0.003 to 0.119 ± 0.030 mm$^3$. The VL of GV was significantly lower ($p < 0.05$) than that of the other three materials evaluated in this study. No significant difference ($p > 0.05$) in VL was found between VE and either PV or RV, although PV and RV were significantly different ($p < 0.05$) from each other. The MDs for materials in this study ranged from 100.439 ± 26.534 to 215.958 ± 27.320 μm. The MD of GV was significantly less ($p < 0.05$) than that of the other three materials evaluated in this study. The MD of PV was significantly greater ($p < 0.05$) than that of the other three materials evaluated in this study. The ADs for materials in this study ranged from 56.053 ± 7.074 to 81.531 ± 7.712 μm. The AD of GC was significantly less ($p < 0.05$) than that of VE, but there were no significant differences between the other materials.

**Table 2.** Results for localized wear of light-cure resin luting cements.

| Materials | Volume Loss (mm$^3$) | Maximum Depth (μm) | Mean Depth (μm) |
|---|---|---|---|
| G-Cem Veneer | 0.027 (0.003) | 100.439 (26.354) | 56.053 (7.074) $^{a}$ |
| Panavia Veneer LC | 0.119 (0.030) $^{a}$ | 215.958 (27.320) | 72.398 (20.853) $^{a,b}$ |
| RelyX Veneer Cement | 0.073 (0.012) $^{b}$ | 147.316 (29.896) $^{a}$ | 71.900 (12.489) $^{a,b,c}$ |
| Variolink Esthetic LC | 0.096 (0.022) $^{a,b}$ | 152.051 (10.910) $^{a}$ | 81.531 (7.712) $^{b,c}$ |

Same lowercase letter in same vertical column indicates no significant difference.

### 3.2. Correlation Analysis

The mean VL of all materials was 0.078 ± 0.039 mm$^3$, the mean MD of all materials was 153.034 ± 47.744 μm and the mean AD of all materials was 70.182 ± 15.371 μm. VL and MD showed a significant positive correlation (R = 0.83, $p < 0.001$). Likewise, VL and MD showed a significant positive correlation (R = 0.47, $p < 0.001$). Furthermore, MD and AD showed a significant positive correlation (R = 0.33, $p < 0.02$) (Figures 3–5).

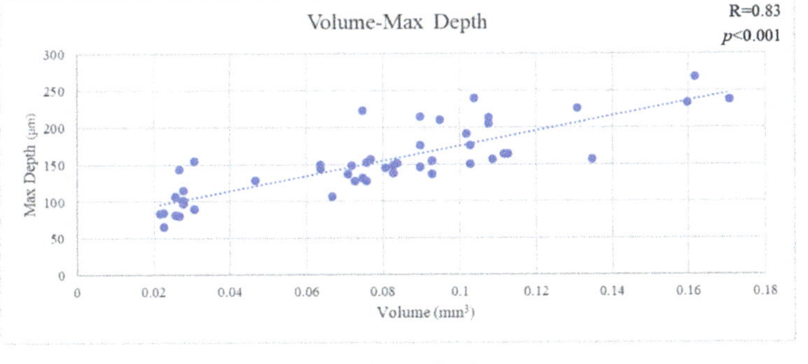


**Figure 3.** The correlation of volume and max depth.

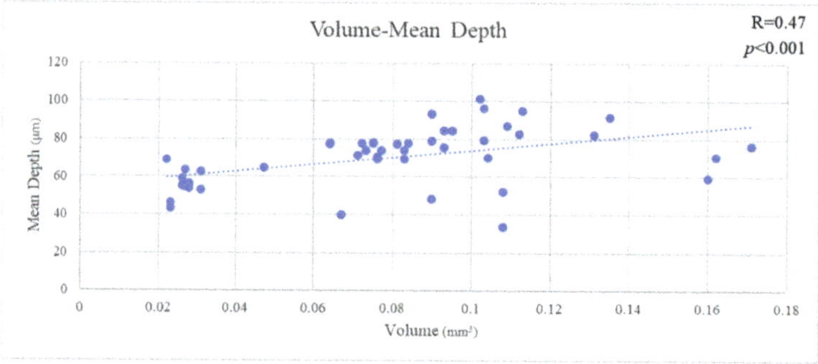


**Figure 4.** The correlation of volume and mean depth.

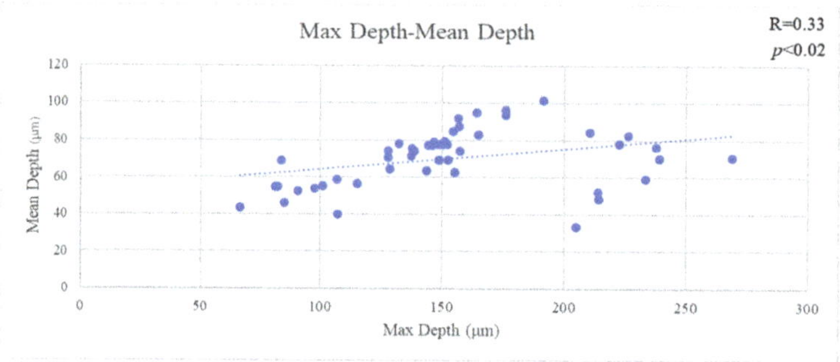


**Figure 5.** The correlation of max depth and mean depth.

### 3.3. Correlation Analysis (Each Material)

The results of the statistical analyses of the various materials are shown in Table 3. A positive correlation was observed for GV, RV and VE in each comparison. However, in PV, only volume–max depth showed a positive correlation, while volume–mean depth and max depth–mean depth showed negative correlations, although these were not significant ($p > 0.05$). For GV and PV, there was a significant positive correlation between volume loss and maximum depth, and no significant correlation between volume loss and mean depth, or between maximum depth and mean depth for PV. For GV and VE, there was a significant positive correlation (r = 0.57 (RV), r = 0.64 (VE)) between maximum depth and mean depth. RV showed no significant correlation (r = 0.26 (max depth), r = 0.44 (mean depth)) between volume loss and either measure of depth, while VE showed a strong positive correlation (r = 0.83) between volume loss and mean depth.

**Table 3.** Results of Pearson correlation analysis for each material.

| Material | Volume<br>Max Depth | | Volume<br>Mean Depth | | Max Depth<br>Mean Depth | |
|---|---|---|---|---|---|---|
| | R | *p*-Value | R | *p*-Value | R | *p*-Value |
| GV | 0.58 | 0.05 | 0.16 | 0.62 | 0.57 | 0.05 |
| PV | 0.73 | <0.01 | −0.10 | 0.76 | −0.35 | 0.26 |
| RV | 0.26 | 0.41 | 0.44 | 0.15 | 0.54 | 0.07 |
| VE | 0.55 | 0.06 | 0.83 | <0.01 | 0.63 | <0.01 |

### 3.4. SEM Observation

Representative SEM images of wear facets after localized wear testing are shown in Figures 6–9. The lower magnification SEM images show clear differences in wear facet size and worn surface characteristics depending on material and also show clear differences in the form of the wear facets, particularly for GV. The higher magnification SEM images clearly show that there were differences in the size and shape of fillers in the light-cure resin luting cement. SEM images of GV showed quite small (<1 μm) irregular particles. SEM images of PV showed irregular and spherical particles with a broad size range (<1 to 5 μm). SEM images of RV and VE showed irregular particles with a broad size range (<1 to 15 μm).

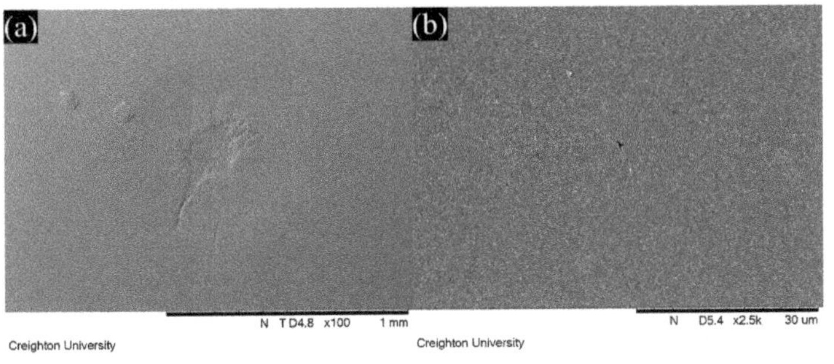


**Figure 6.** SEM images of the wear facets of G-Cem Veneer as viewed at (**a**) 100× and (**b**) 2500× magnification.

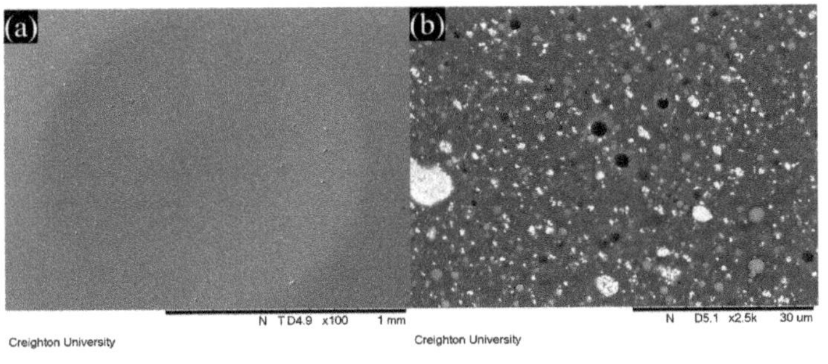


**Figure 7.** SEM images of the wear facets of Panavia V5 LC as viewed at (**a**) 100× and (**b**) 2500× magnification.

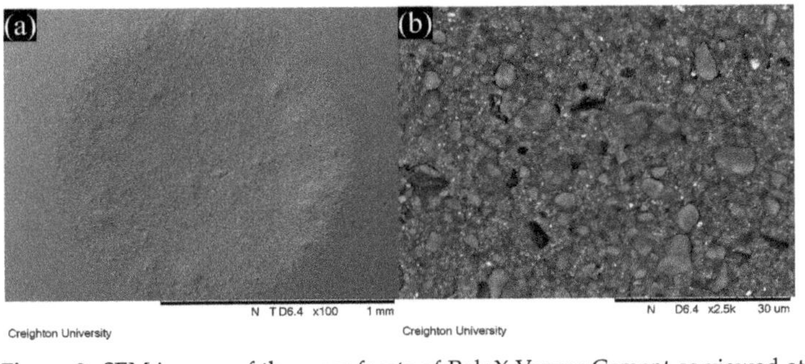


**Figure 8.** SEM images of the wear facets of RelyX Veneer Cement as viewed at (**a**) 100× and (**b**) 2500× magnification

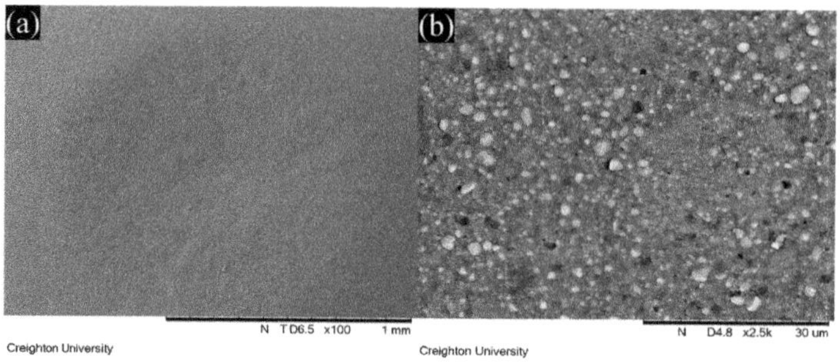


**Figure 9.** SEM images of the wear facets of Variolink Esthetic LC as viewed at (**a**) 100× and (**b**) 2500× magnification.

## 4. Discussion

Our previous study reported that the localized wear of dual-cure resin luting cements had a volume loss of 0.078–0.126 $mm^3$ and a maximum depth of 152.2–187.4 μm [16]. Some clinicians like to use dual-cure resin luting cements due to their versatile usage even with their known increase in susceptibility to discoloration. However, in the results of this study, most of the light-cured resin luting cements tested showed wear susceptibility within the range measured for dual-cure cements, and GV was clearly superior in this respect. This suggests that although light-cure resin luting cements may be a better option, when there is no problem with light irradiation of the material, the careful choice of the cement is still important to achieve more favorable results.

Flowable resin-based composites (RBCs) are sometimes used as a substitute for light-cure resin luting cements, because of concerns about the aesthetic quality and wear resistance of cement that is exposed at the margins of a restoration. Our previous study on the wear resistance of flowable RBCs found values for volume loss of 0.025–0.148 $mm^3$ and maximum depths of 98.1–210.6 μm [17]. These values are not significantly different from the range found in this study. Our recent study of paste-type RBCs found values for volume loss of 0.034–0.059 $mm^3$ and a maximum depth of 110.6–151.5 μm [18]. These values are generally superior to those of light-cured resin luting cements, but in this study GV showed superior wear resistance to that found for paste-type RBCs in the past. These results suggest that wear resistance is not a good reason to use an RBC instead of a resin luting cement, particularly if the RBC must be heated, and that light-cured cements can be used in cases where adequate irradiation can be achieved, such as with ceramic veneers.

On the other hand, the wide range of wear values indicates that there are substantial differences between different materials. This was also observed in this study, in which GV showed superior wear resistance to the other materials. The manufacturer claims that this material uses 150 nm diameter filler particles, which are smaller than those used in other resin luting cements [18]. This claim is supported by the SEM images obtained in this study, which show smaller filler particles for GV than for the other materials investigated. It is possible that the size of the filler particles contributes to the improved wear resistance.

If the size of the filler particles is influencing the wear resistance, it would also be expected to affect the shape of the wear facets, and this can be seen in the SEM images. It is normally assumed that a typical wear facet is roughly hemispherical, with its deepest point at the center and a uniform reduction in depth towards the edges. However, it is clear both from the images and the correlations between the various measurements that things are not so simple. An irregular depression is clearly visible within the wear facet of GV, and that of PV appears to include an elevated region near the center. The fact that the overall correlation between the maximum depth and the mean depth is only 0.33 shows that the facets cannot, in general, be regularly shaped. Further, neither the mean depth nor the maximum depth is correlated with volume loss for all the materials, and some do show a statistically significant correlation for three of the four, in two cases the correlation is with maximum depth, and in one case it is with mean depth. One material (RV) shows no statistically significant correlation between volume loss and either measurement of depth ($p > 0.05$).

This lack of correlation strongly suggests that the wear facets are irregular in shape, and that this irregularity differs from one material to another. As the SEM images show clear differences in the size and shape of the fillers, it is reasonable to hypothesize that these differences are due to filler loss at the surface of the wear facet leading to irregular depressions in the wear surface. This hypothesis is qualitatively consistent with the SEM images of the wear facets, although the irregular depression visible for GV does not appear to have this cause. These results indicate that it is important to use multiple measures of localized wear in order to accurately understand the wear properties of a material, because the correlations between the ways of measuring wear are often weak or non-existent and may vary depending on the material.

In this study, GV showed superior wear resistance to the other materials, and a different wear pattern was observed in the SEM imaging. GV uses smaller filler particles than the other materials, which could plausibly improve wear resistance by reducing the exposure of softer resin matrix at the surface. However, as this study only considered one material with smaller particles, other factors cannot be ruled out.

The main limit of this study is that the cements were exposed to wear in isolation from ceramics and tooth material, a situation that is never found in clinical cases. However, an earlier study compared results for the wear resistance of blocks of resin luting cement and narrow samples between other materials, and found a linear relationship, which suggests that this method is suitable for the comparison of materials [5].

The small number of materials makes it difficult to say anything definite about the causes of the observed differences in wear resistance, but there are few materials of this sort on the market, and so a wide comparison is not possible.

The ideal is for clinicians to choose the best resin luting cement for the situation of each individual patient. However, this requires the clinic to keep a wide range of different materials on hand, and is often impractical. As a result, there is a trend towards the development of universal self-adhesive resin luting cements, which may be used with primer or universal adhesive if necessary to achieve stronger bonding [16]. It is therefore important to also assess the wear resistance of these materials, and compare them to existing dual-cure and light-cure resin luting cements, to determine whether it is appropriate to use them in clinical contexts.

It would also be valuable to perform these tests on experimental versions of these cements, in which the proportion and size of filler or composition of the resin can be varied while keeping other factors constant. This would enable a more definite conclusion on the causes of superior wear resistance, and might contribute to the development of better materials for the future. The results of this study suggest that filler size might be an important factor to vary in such research.

## 5. Conclusions

The light-cure resin luting cements investigated in this study generally show similar wear resistance to dual-cure cements, and also to flowable resin-based composites, but certain cements show superior wear resistance to possible alternative materials, including most paste-type resin-based composites. This class of materials may be suitable for consideration for clinical use from a wear perspective.

**Author Contributions:** Conceptualization, T.K. and A.T.; methodology, T.K. and A.T.; software, T.K., T.H. and A.T.; validation, M.O., T.K., T.H. and A.T.; formal analysis, T.K., T.H. and A.T.; investigation, A.T.; resources, W.W.B. and A.T.; data curation, A.T.; writing—original draft preparation, T.K. and A.T.; writing—review and editing, A.A., M.I., W.W.B. and A.T.; visualization, M.O., T.K. and V.C.S.; supervision, W.W.B. and A.T.; project administration, W.W.B. and A.T.; funding acquisition, A.A. All authors have read and agreed to the published version of the manuscript.

**Funding:** This research was supported by a grant from Researchers Supporting Project number (RSPD2024R790), King Saud University, Riyadh, Saudi Arabia.

**Institutional Review Board Statement:** Not applicable.

**Informed Consent Statement:** Not applicable.

**Data Availability Statement:** The original contributions presented in the study are included in the article, further inquiries can be directed to the corresponding author.

**Acknowledgments:** The authors thank the grant from the Researchers Supporting Project number (RSPD2024R790), King Saud University, Riyadh, Saudi Arabia.

**Conflicts of Interest:** The funding sponsors had no role in the design of the study; in the collection, analyses, or interpretation of data; in the writing of the manuscript, or in the decision to publish the results.

## References

1. Morimoto, S.; Albanesi, R.B.; Sesma, N.; Agra, C.M.; Braga, M.M. Main clinical outcomes of feldspathic porcelain and glass-ceramic laminate veneers: A systematic review and meta-analysis of survival and complication rates. *Int. J. Prosthodont.* **2016**, *29*, 38–49. [CrossRef] [PubMed]
2. Ghodsi, S.; Shekarian, M.; Aghamohseni, M.M.; Rasaeipour, S.; Arzani, S. Resin cement selection for different types of fixed partial coverage restorations: A narrative systematic review. *Clin. Exp. Dent. Res.* **2023**, *9*, 1096–1111. [CrossRef] [PubMed]
3. Beier, U.S.; Kapferer, I.; Burtscher, D.; Dumfahrt, H. Clinical performance of porcelain laminate veneers for up to 20 years. *Int. J. Prosthodont.* **2012**, *25*, 79–85.
4. Aslan, Y.U.; Uludamar, A.; Özkan, Y. Clinical performance of pressable glass-ceramic veneers after 5, 10, 15, and 20 years: A retrospective case series study. *J. Esthet. Restor. Dent.* **2019**, *31*, 415–422. [CrossRef]
5. Tsujimoto, A.; Barkmeier, W.W.; Takamizawa, T.; Latta, M.A.; Miyazaki, M. Relationship between simulated gap wear and generalized wear of resin luting cements. *Oper. Dent.* **2017**, *42*, e148–e158. [CrossRef]
6. Turssi, C.P.; De Moraes Purquerio, B.; Serra, M.C. Wear of dental resin composites: Insights into underlying processes and assessment methods—A review. *J. Biomed. Mater. Res. B Appl. Biomater.* **2003**, *65*, 280–285. [CrossRef] [PubMed]
7. Blunck, U.; Fischer, S.; Hajtó, J.; Frei, S.; Frankenberger, R. Ceramic laminate veneers: Effect of preparation design and ceramic thickness on fracture resistance and marginal quality in vitro. *Clin. Oral Investig.* **2020**, *24*, 2745–2754. [CrossRef]
8. Al-Akhali, M.; Chaar, M.S.; Elsayed, A.; Samran, A.; Kern, M. Fracture resistance of ceramic and polymer-based occlusal veneer restorations. *J. Mech. Behav. Biomed. Mater.* **2017**, *74*, 245–250. [CrossRef]
9. Stappert, C.F.; Abe, P.; Kurths, V.; Gerds, T.; Strub, J.R. Masticatory fatigue, fracture resistance, and marginal discrepancy of ceramic partial crowns with and without coverage of compromised cusps. *J. Adhes. Dent.* **2008**, *10*, 41–48. [PubMed]
10. Hardan, L.; Bourgi, R.; Hernández-Escamilla, T.; Piva, E.; Devoto, W.; Lukomska-Szymanska, M.; Cuevas-Suárez, C.E. Color stability of dual-cured and light-cured resin cements: A systematic review and meta-analysis of in vitro studies. *J. Prosthet. Dent.* **2024**, *33*, 212–220. [CrossRef]
11. Yadav, R.; Meena, A.; Lee, H.H.; Lee, S.Y.; Park, S.J. Tribological behavior of dental resin composites: A comprehensive review. *Tribol. Int.* **2023**, *190*, 109017. [CrossRef]
12. Spazzin, A.O.; Guarda, G.B.; Oliveira-Ogliari, A.; Leal, F.B.; Correr-Sobrinho, L.; Moraes, R.R. Strengthening of porcelain provided by resin cements and flowable composites. *Oper. Dent.* **2016**, *41*, 179–188. [CrossRef]
13. Shinkai, K.; Suzuki, S. Effect of cyclic impact loading on the surface properties of flowable resin composites. *Dent. Mater. J.* **2014**, *33*, 874–879. [CrossRef] [PubMed]
14. Prieto, L.T.; Souza, E.J., Jr.; Araújo, C.T.; Lima, A.F.; Dias, C.T.; Paulillo, L.A. Knoop hardness and effectiveness of dual-cured luting systems and flowable resin to bond leucite-reinforced ceramic to enamel. *J. Prosthodont.* **2013**, *22*, 54–58. [CrossRef]

15. Barbon, F.J.; Isolan, C.P.; Soares, L.D.; Bona, A.D.; de Oliveira da Rosa, W.L.; Boscato, N. A systematic review and meta-analysis on using preheated resin composites as luting agents for indirect restorations. *Clin. Oral. Investig.* **2022**, *26*, 3383–3393. [CrossRef] [PubMed]
16. Tsujimoto, A.; Barkmeier, W.W.; Takamizawa, T.; Watanabe, H.; Johnson, W.W.; Latta, M.A.; Miyazaki, M. Simulated localized wear of resin luting cements for universal adhesive systems with different curing mode. *J. Oral Sci.* **2018**, *60*, 29–36. [CrossRef]
17. Tsujimoto, A.; Barkmeier, W.W.; Takamizawa, T.; Latta, M.A.; Miyazaki, M. Influence of thermal stress on simulated localized and generalized wear of nanofilled resin composites. *Oper. Dent.* **2018**, *43*, 380–390. [CrossRef] [PubMed]
18. Ujiie, M.; Tsujimoto, A.; Barkmeier, W.W.; Jurado, C.A.; Villalobos-Tinoco, J.; Takamizawa, T.; Latta, M.A.; Miyazaki, M. Comparison of occlusal wear between bulk-fill and conventional flowable resin composites. *Am. J. Dent.* **2020**, *33*, 74–78. [PubMed]

Journal of
*Functional Biomaterials*

MDPI

*Systematic Review*

# Advanced Platelet-Rich Fibrin Plus (A-PRF+) as an Additive to Hard Tissue Managing Protocols in Oral Surgery: A Systematic Review

**Marek Chmielewski [1,*], Andrea Pilloni [2] and Paulina Adamska [3,*]**

[1] Private Dental Practice, 14 Kolberga Street, 81-881 Sopot, Poland
[2] Section of Periodontics, Department of Oral and Maxillo-Facial Sciences, Sapienza Unviersity of Rome, 00-185 Rome, Italy; andrea.pilloni@uniroma1.it
[3] Department of Oral Surgery, Medical University of Gdańsk, 7 Dębinki Street, 80-211 Gdańsk, Poland
* Correspondence: m.chmielewski@gumed.edu.pl (M.C.); paulina.adamska@gumed.edu.pl (P.A.)

**Abstract:** Background: Advanced platelet-rich fibrin + (A-PRF+) represents a third generation of autologous platelet derivatives. Appropriate centrifugation conditions cause the formation of a clot containing platelets, which slowly release growth factors that influence healing. The objective of this article was to undertake a review of the available literature on the effectiveness of A-PRF+ use in hard tissue procedures. Materials and methods: In order to ensure the most accurate and relevant results, only randomized clinical trials regarding bone regeneration techniques/bone healing that compared the effect of the A-PRF+ addition in dentistry were included in this study. Articles taken into consideration for the review were published between the beginning of 2014 and 31 December 2024. The search of manuscripts for the review was conducted using the PubMed, Scopus, Google Scholar, and Cochrane databases. For this study, 10 articles focusing on A-PRF+ were qualified. Results: A-PRF+ was found to increase the post-surgical vertical and horizontal alveolar ridge dimensions. The bone formed in the surgical site presented a higher volume of vital and non-vital bone and a more optimal bone composition, at the same time providing a lower percentage of connective tissue inclusions. When combined with other grafting biomaterials, A-PRF+ enhanced their performance and integration. A-PRF+ did not have any significant effect on the mineral bone density compared with other grafting materials. Compared with PRF and other blood derived plasmas rich in growth factors, the performance of A-PRF+ was generally better, but often with no statistical significance. The treatment of periodontal defects measured by the reduction in pocket depth and clinical attachment level also fared better with the A-PRF+ addition, although there was no differences noted between A-PRF+ and biphasic calcium phosphate and xenograft. Finally, the A-PRF+ addition improved the primary implant stability in the evaluated studies. Conclusions: The A-PRF+ addition to the surgical protocols significantly enhanced the healing of the bone and when combined with biomaterials improved their integration and increased the implant insertion torque, improving the primary and secondary stability. It may be a viable alternative for patients that express their concern towards human- and animal-derived biomaterials.

**Keywords:** advanced platelet-rich fibrin; A-PRF; autografts; dentistry; growth factors; wound healing

## 1. Introduction

Dr. Jospeh Choukroun is widely considered as a promoter of the use of blood-derived platelet-rich biomaterials rich in growth factors. Choukroun, together with his team, developed and distributed the first protocols for preparation at the beginning of the 21st century. Studies on blood-derived, platelet-rich preparations enriched with growth factors have been conducted to improve their biological and physical properties. Researchers aimed to develop preparations with greater cohesion, allowing for better positioning, particularly in bone defects during procedures in oral surgery, periodontology, and implantology, for example, after tooth extraction and in bone regeneration (bone augmentation, sinus lift procedures, treatment periodontitis, or during implant placement). Additionally, it was found that the new generations of these preparations contain novel growth factors and cytokines, which enhance their biological properties. This leads to reduced post-procedure reactions (such as pain and swelling), faster wound healing, and other benefits [1–4].

Available studies on blood-derived, platelet-rich preparations enriched with growth factors emphasize their role in enhancing healing and reducing procedural discomfort, particularly during the first days or up to two weeks after the procedure. This applies to research on platelet-rich plasma as well as various types of platelet-rich fibrins. This period is crucial for patient recovery, and minimizing discomfort and potential complications significantly improves the patient comfort and quality of life post-procedure. Additionally, combining these preparations with bone-derived and bone substitute materials has been shown to enhance healing outcomes, such as by increasing the effectiveness of bone augmentation, sinus lift procedures, and implant treatments. Therefore, incorporating them into treatment protocols for oral procedures may be a rational approach [4–16].

The advanced platelet-rich fibrin (A-PRF) manufacturing method was first introduced and described by Ghanaati et al. [5] and Choukroun [6] in 2014, which was further refined by Kobayashi to create A-PRF+ [7]. A-PRF is a biocompatible material obtained from the patient's venous blood. The standard procedure for the production of A-PRF+ entails the collection of the peripheral venous blood of the patient and then the utilization of vertical centrifuging at a rotational speed of 1300–1500 rpm for a period of 14 min, as outlined by Kobayashi. A further variant, designated A-PRF+, has been developed, which requires a different centrifuging protocol of 1300 rpm for 8 min and was designed to further enhance the standard A-PRF protocol [7]. The vertical centrifugation process facilitates the separation and sedimentation of cellular components according to their respective molecular masses, as proposed by Herrera-Vizcaíno [8]. Moreover, a reduction in the centrifuge speed compared to the standard PRF procedure, which involves speeds between 2700–3000 rpm for a duration of 12 min, allows for a superior release of growth factors. It has been demonstrated that platelets remain present in the peripheries of the A-PRF+ clot. The discrepancy in processing may be accountable for the enhanced optimization, longevity, and more uniform distribution and liberation of growth factors from A-PRF into adjacent tissues, thereby influencing the regeneration and maturation of these tissues. The lymphocytes, macrophages, and stem cells are concentrated in the proximal part of the clot, whereas neutrophils are located mainly in the distal part [5,6,8]. A-PRF and A-PRF+ have been proven to contain cytokines and growth factors such as the following: bone morphogenetic proteins, fibroblast growth factor, matrix metalloproteinases, platelet-derived growth factor, transforming growth factors $\alpha$ and $\beta$, and vascular endothelial growth factor [16–20]. Due to the modified centrifugation parameters, A-PRF+ contains a higher concentration of growth factors and cytokines compared to A-PRF. This enhances its angiogenic properties and may improve tissue regeneration during healing. Additionally, its higher density increases the stability and retention in the wound compared to A-PRF. The A-PRF+ mesh traps more growth factors and neutrophils. This ensures good cohesion and attachment to the wound. The available evidence suggests that A-PRF (and its more recent iteration, A-PRF+) can release cytokines for up to 10 days. A-PRF+ is an additional tool that can be used to improve the result of the standard surgery procedure [6,9–12].

The objective of this article was to undertake a review of the available literature on the effectiveness of A-PRF+ use in hard tissue management procedures. This article provides a clear and structured summary of the available publications on the improved A-PRF formula and its role in oral surgery, periodontology, and implantology.

## 2. Materials and Methods

The review was carried out beginning with the search of the literature, which was conducted using PubMed, Scopus, Google Scholar, and Cochrane web databases to answer the questions: 'Does A-PRF+ addition improve integration of bone grafting materials in oral surgery procedures?' and 'Does A-PRF+ addition improve hard tissue healing in oral surgery procedures?' To search for relevant publications, the following MeSH terms were used: 'platelet-rich fibrin', 'PRF', 'autografts', 'dentistry', 'growth factors', and 'wound healing'. The search was conducted for manuscripts published between 1 January 2014 and 31 December 2024.

This project was registered in the International Prospective Register of Systematic Reviews (PROSPERO) and granted the number CRD42023449648.

### 2.1. Inclusion and Exclusion Criteria

In order to ensure the most accurate and relevant results, only randomized clinical trials regarding bone regeneration techniques that compared the effect of A-PRF+ addition were included in this study. The studies were selected from the databases according to the following inclusion criteria: (1) only studies conducted on human subjects were included, (2) studies that employed A-PRF+ in conjunction with biomaterials for bone regeneration techniques, (3) studies that were carried out and published between the 1 January 2014 and 31 December 2024, (4) studies that were published in English, and (5) randomized clinical studies that included a minimum of 10 patients.

The exclusion criteria included the following: (1) studies not conducted on human subjects, (2) studies that employed A-PRF (not A-PRF+) in conjunction with biomaterials for bone regeneration techniques, (3) studies that employed A-PRF+ as an additive in managing soft tissue surgical protocols, (4) studies that were carried out and published before 2014 (before the development of A-PRF+), (5) studies that were not published in English, and (6) case reports and articles that included a less than 10 patients.

### 2.2. Screening and Data Extraction

Two reviewers (MC and PA) independently performed the data extraction. Any disagreements during the extraction process were resolved by discussion between the reviewers. Duplicates from databases were excluded. Lastly, the full-text manuscripts were subjected to a review in accordance with the established selection criteria. A total of 10 publications were deemed eligible for inclusion in this review.

## 3. Results

In the initial stages of the selection process, a total of 217 references were identified through searches of the PubMed, Scopus, Google Scholar, and Cochrane databases. Following the removal of duplicates, 114 articles were retained. Following the screening of the titles and abstracts, 140 positions were excluded. Full texts were then read, and only publications that were randomized trials about A-PRF+ were included. Case reports, reviews, or articles about A-PRF/PRF/PRP (not specifically about A-PRF+) were excluded.

Ultimately, 10 articles were deemed suitable for inclusion in this systematic review [13–22]. Subsequently, the remaining studies were divided into the appropriate categories according to the procedures they employed. The initial studies included in the review were published in 2015 (Figure 1 and Table 1).

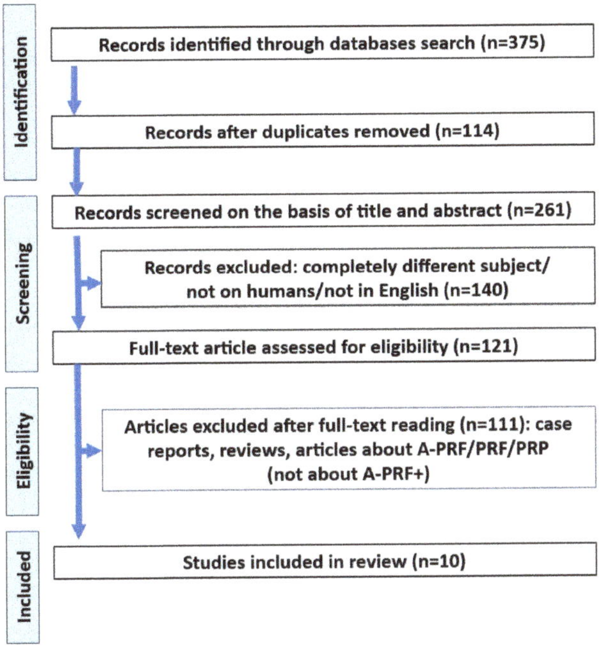


**Figure 1.** PRISMA flowchart.

**Table 1.** Studies included in review.

| No. | References | Country | Number of Patients | Preparation of A-PRF |
|---|---|---|---|---|
| 1 | Alhaj et al., 2018 [15] | Lebanon | 20 | 1300 rpm for 8 min (A-PRF+) |
| 2 | Angelo et al., 2015 [22] | Austria | 82 | A-PRF+ (exact preparation not specified) |
| 3 | Clark et al., 2018 [13] | USA | 40 (divided into 4 groups) | 1300 rpm for 8 min (A-PRF+) |
| 4 | Dragonas et al., 2023 [19] | USA | 15 | 1300 rpm for 5 and 3 min (8 min total) with liquid layer removal after 5 min centrifuge (A-PRF+ modified protocol) |
| 5 | Ghonima et al., 2020 [21] | Egypt | 22 | 1300 rpm for 8 min (A-PRF+) |
| 6 | Hartlev et al., 2020 [18] | Denmark | 27 (divided into 2 groups) | 1300 rpm for 14 min (A-PRF+ modified) |
| 7 | Ivanova et al., 2019 [14] | Bulgaria | 60 (divided into 3 groups) | 1300 rpm for 8 min (A-PRF+) |
| 8 | Kalash et al., 2017 [17] | Lebanon | 18 | 1300 rpm for 8 min (A-PRF+) |
| 9 | Lavagen et al., 2021 [20] | France | 24 | 1300 rpm for 5 and 3 min (8 min total) with liquid layer removal after 5 min centrifuge (A-PRF+ modified protocol) |
| 10 | Yewale et al., 2021 [16] | India | 20 | 1300 rpm for 8 min (A-PRF+) |

### 3.1. A-PRF+ Effect on Vertical and Horizontal Alveolar Ridge Dimensions

A total of five randomized clinical trials were conducted to evaluate the alveolar ridge dimensions following tooth extractions and grafting procedures. In four out of the five studies, the incorporation of A-PRF+ into the surgical protocol resulted in an improvement in the clinical outcome or facilitated the use of co-used biomaterials, thereby producing more favorable results [13–17].

In the study conducted by Clark et al. [13], the blood clot was employed as a control. The test groups employed a freeze-dried bone allograft (FDBA), A-PRF+, or a combination of the two materials. Each test group demonstrated a superior performance in maintaining three-dimensional bone measurements in comparison to the control group. When employed independently, A-PRF+ demonstrated comparable outcomes to the FDBA with respect to the vertical and horizontal bone loss. The incorporation of A-PRF+ into the FDBA resulted in a notable enhancement in the vertical measurements. The incorporation of biomaterials did not exert a notable impact on the horizontal dimensions in comparison to the control group at the apical region of the alveolar ridge. However, the test group, comprising the FDBA and A-PRF+, demonstrated a significantly superior capacity to impede the loss of the coronal horizontal dimensions in comparison to all other groups.

A comparable trial was conducted by Ivanova et al. [14], in which the test group that had used the mixture of FDBA and A-PRF+ was omitted. In the study, A-PRF+ demonstrated a comparable performance to the FDBA. The results for both the FDBA and A-PRF+ groups in terms of the vertical and horizontal dimensions were statistically significant when compared to those of the blood clot control group. Both the Clark and Ivanova studies had approximately the same follow-up time of four months. The follow-up time in the Clark study was 3.75 months, while in the Ivanova study it was 4 months.

In their respective studies, Alhaj et al. [15] and Yewale et al. [16] employed A-PRF+ as an adjunct to the bone biomaterial. A minimal discrepancy between the groups was observed on the second day of the study by Alhaj et al. [15]. At the six-month mark, the control and test groups exhibited statistically significant differences in favor of the test group. Furthermore, Yewale et al. [16] observed that the addition of A-PRF+ enhanced the performance of other hard tissue grafting materials. The most significant gains were observed in the horizontal dimensions, particularly at the 3 mm and 5 mm measuring points.

Kalash et al. [17] employed an xenograft material with (test) and without (control) A-PRF+. The results of this study differ from those of the others, as the resorption margin in the test group was found to be higher than in the control group. No statistically significant difference was observed between the two groups. The results are presented in Table 2 [13–17].

**Table 2.** Differences in vertical and horizontal dimensions compared to control and other materials.

| No. | References | Vertical Dimensions Loss (mm) | | | | | | Horizontal Dimensions (mm) | | | | | |
|---|---|---|---|---|---|---|---|---|---|---|---|---|---|
| 1 | Alhaj et al. [15] | Marginal bone height | Autologous bone graft | Autologous bone graft + A-PRF+ | | | *p* value | No data | | | | | |
| | | Day 2 | 0.515 | 0.435 | | | 0.009 | | | | | | |
| | | 6 months | 0.205 | 0.35 | | | <0.001 | | | | | | |
| 2 | Clark et al. [13] | 15 weeks | Blood clot | A-PRF+ | FDBA + A-PRF+ (1:1 ratio) | FDBA | *p* value | 15 weeks | Blood clot | A-PRF+ | FDBA + A-PRF+ (1:1 ratio) | FDBA | *p* value |
| | | | 3.8 ± 2 | 1.8 ± 2.1 | 1.0 ± 2.3 | 2.2 ± 1.8 | Blood clot vs. A-PRF+ and FDBA + A-PRF+ <0.05 | Coronal | 2.9 ± 1.7 | 2.8 ± 1.2 | 1.9 ± 1.1 | 2.5 ± 1.1 | FDBA + A-PRF+ < 0.05 vs. other group |
| | | | | | | | | Middle | 1.8 ± 1.3 | 1.8 ± 1.8 | 1.7 ± 1.2 | 1.5 ± 1.2 | |
| | | | | | | | | Apical | 1.6 ± 1.5 | 1.8 ± 1.5 | 1.6 ± 1.5 | 1.2 ± 1.3 | |
| 3 | Ivanova et al. [14] | | Blood clot | A-PRF+ | | FDBA | *p* value | | Blood clot | A-PRF+ | | FDBA | *p* value |
| | | 4 months | 1.38 ± 0.13 | 0.87 ± 0.21 | | 0.91 ± 0.24 | A-PRF+ and FDBA > 0.05; blood clot < 0.05 | 4 months | 2.39 ± 0.4 | 1.52 ± 0.31 | | 1.33 ± 0.25 | A-PRF+ and FDBA > 0.05; blood clot < 0.05 |
| 4 | Kalash et al. [17] | Marginal bone height | Xenograft | Xenograft + A-PRF | | | *p* value | No data | | | | | |
| | | Baseline | 1.79 ± 0.2 | 1.91 ± 0.22 | | | | | | | | | |
| | | 9 months | 1.3 ± 0.32 | 1.37 ± 0.2 | | | | | | | | | |
| 5 | Yewale et al. [16] | 6 months | Sybograf plus (70% HA, 30% BTCP) + A-PRF+: | Sybograf plus (70% HA, 30% BTCP | | | *p* value | 6 months | Sybograf plus (70% HA, 30% BTCP) + A-PRF+ | Sybograf plus (70% HA, 30% BTCP | | | *p* value |
| | | | | | | | | Mean difference | Socket width at 1 mm | | | | |
| | | | | | | | | | 2.12 | 1.83 | | | 0.49 |
| | | | 1.4 | 1.67 | | | <0.05 | | Socket width at 3 mm | | | | |
| | | | | | | | | | 1.68 | 0.59 | | | 0.041 |
| | | | | | | | | | Socket width at 5 mm | | | | |
| | | | | | | | | | 0.97 | 0.33 | | | 0.65 |

A-PRF+—advanced platelet-rich fibrin +; BTCP—β-tricalcium phosphate; FDBA—freeze-dried bone allograft; and HA—hydroxyapatite.

### 3.2. A-PRF+'s Effect on Vital Bone Formation, Grafting Material Turnover, and Percentage of Connective Tissue in Grafting Site

The grafting material integration, the turnover, and the formation of bone and connective tissue were investigated in five randomized controlled trials. The incorporation of A-PRF+ into the protocol resulted in an enhanced tissue formation when compared with the control groups and facilitated a more effective redistribution of the grafting material at the surgical sites [13,14,18–20].

All of the studies considered reported on the formation of the bone tissue. In the study conducted by Clark et al. [13], the incorporation of a single biomaterial resulted in a reduction in the vital bone formation when compared to the other experimental groups (29% vs. 40–46%). However, when the same biomaterial was combined with A-PRF+, the percentage of vital bone increased to a level comparable to that observed in the control group. Furthermore, the incorporation of A-PRF+ resulted in a notable reduction in the connective tissue percentage; however, in the Clark et al. study, this decline was relatively minor. Similarly, Ivanova observed a favorable impact of the A-PRF+ supplementation on the bone vitality in comparison to the control group. No statistically significant difference was observed between the two test groups. Furthermore, the addition of A-PRF+ has been observed to significantly influence the quantity of the connective tissue present within the harvested specimen. The addition of A-PRF+ had a comparable effect to the FDBA, which inhibited the formation of connective tissue and promoted bone regeneration [13]. Additionally, Hartlev et al. [18] investigated the quantity of non-vital bone. A comparison of the A-PRF+ and deproteinized bovine bone mineral (DBBM) groups revealed comparable levels of vital bone. The two groups exhibited notable differences in the non-vital bone percentage (A-PRF+ 80%, DBBM 63%) and the percentage of connective tissue (5% vs. 22%). In comparison to the DBBM, the A-PRF+ specimens exhibited a greater proportion of non-vital bone, accompanied by a reduction in the connective tissue within the bone structure. Dragonas et al. [19] and Lavagen et al. [20] also employed the DBBM grafting material in their study, yet no differentiation was made between vital and non-vital bone. The addition of A-PRF+ in Dragonas et al. [19]'s study resulted in a greater extent of bone formation than that observed in the sole DBBM control group (20.33 $\pm$ 11.5% vs. 32.2 $\pm$ 7.29%), although this was similar to what was observed in another test group that used plasma rich in growth factors (PRGF) (32.2 $\pm$ 7.29% vs. 34.8 $\pm$ 6.83%). The increase in the bone formation observed in both test groups in comparison to the control group was statistically significant. The turnover of the material in the A-PRF+ group was slower than in the control group (24 $\pm$ 7.94% vs. 26 $\pm$ 7.78%) and in the PRGF group (24 $\pm$ 7.94% vs. 15.8 $\pm$ 8.23%). The percentage of connective tissue in the A-PRF+ group was found to be lower than that of the control group (41.4 $\pm$ 8.32% vs. 55.66 $\pm$ 7.77%) and the PRGF group (41.4 $\pm$ 8.32% vs. 49.6 $\pm$ 5.68%). The study conducted by Lavagen et al. [20] concentrated exclusively on the percentage of newly formed bone. In contrast with the findings of other studies, both the test and control groups employed the use of a blood clot derivative. The control group employed an older generation of PRF, whereas the test group utilized A-PRF+. The test group exhibited superior outcomes compared to the control group, with a statistically significant difference in the formation of new bone, reaching 60.4 $\pm$ 10.4% compared to 51.4 $\pm$ 18.4% for the control group. The mean volume of the newly formed bone for the test group was 0.29 $\pm$ 0.09 $cm^3$, while that for the control group was 0.2 $\pm$ 0.08 $cm^3$, indicating that the gain in the A-PRF+ group was statistically significant.

In the study by Ivanova et al. [14], there was no statistical difference between the allograft and A-PRF+ for the vital bone formation. The results were statistically better compared to the control group. The same observations were made in the case of the connective tissue graft. The results are presented in Table 3 [13,14,18–20].

**Table 3.** Differences in vital bone, non-vital bone, and connective tissue formation and residual grafting material (residual material does not apply to A-PRF+, only FDBA and DBBM).

<table>
<tr><th>No.</th><th>References</th><th colspan="5">Vital Bone (%)</th><th colspan="3">Non-Vital Bone (%)</th><th colspan="5">Residual Material (%)</th><th colspan="5">Connective Tissue (%)</th></tr>
<tr><td rowspan="2">1</td><td rowspan="2">Clark et al. [13]</td><td>Blood clot</td><td>A-PRF+</td><td>A-PRF+ and FDBA (1:1 ratio)</td><td>FDBA</td><td>p value</td><td colspan="3" rowspan="2">No data</td><td>Blood clot</td><td>A-PRF+</td><td>A-PRF+ and FDBA (1:1 ratio)</td><td>FDBA</td><td>p value</td><td>Blood clot</td><td>A-PRF+</td><td>A-PRF+ and FDBA (1:1 ratio)</td><td>FDBA</td><td>p value</td></tr>
<tr><td>40 ± 18</td><td>46 ± 18</td><td>40 ± 15</td><td>29 ± 14</td><td><0.05 (for A-PRF group)</td><td>0</td><td>0</td><td>3 ± 3</td><td>11 ± 9</td><td>>0.05</td><td>60 ± 10</td><td>55 ± 15</td><td>58 ± 5</td><td>58 ± 10</td><td>>0.05</td></tr>
<tr><td rowspan="2">2</td><td rowspan="2">Dragonas et al. [19]</td><td colspan="3">Control (DBBM)</td><td colspan="2">Test (A-PRF+ + DBBM)</td><td colspan="2">Test (PRGF + DBBM)</td><td>p value</td><td colspan="2">Control (DBBM)</td><td>Test (A-PRF+ + DBBM)</td><td>Test (PRGF + DBBM)</td><td>p value</td><td>Control (DBBM)</td><td>Test (A-PRF+ + DBBM)</td><td colspan="2">Test (PRGF + DBBM)</td><td>p value</td></tr>
<tr><td colspan="3">20.33 ± 11.5</td><td colspan="2">32.2 ± 7.9</td><td colspan="2">34.8 ± 6.83</td><td>0.0875</td><td colspan="2">24 ± 7.94</td><td>26 ± 7.78</td><td>15.8 ± 8.23</td><td>0.161</td><td>55.66 ± 7.77</td><td>41.4 ± 8.32</td><td colspan="2">49.6 ± 5.68</td><td>0.0573</td></tr>
<tr><td rowspan="2">3</td><td rowspan="2">Hartlev et al. [18]</td><td colspan="2">A-PRF</td><td colspan="2">DBBM</td><td>p value</td><td>A-PRF</td><td>DBBM</td><td>p value</td><td colspan="5" rowspan="2">No data</td><td colspan="2">A-PRF</td><td colspan="2">DBBM</td><td>p value</td></tr>
<tr><td colspan="2">13.75 ± 13.18</td><td colspan="2">14.16 ± 10.11</td><td>1.00</td><td>80.06 ± 15.28</td><td>63.16 ± 29.81</td><td>0.19</td><td colspan="2">4.89 ± 6.05</td><td colspan="2">21.96 ± 21.96</td><td>0.11</td></tr>
<tr><td rowspan="2">4</td><td rowspan="2">Ivanova et al. [14]</td><td colspan="2">Blood clot</td><td>A-PRF</td><td>FDBA</td><td>p value</td><td colspan="3" rowspan="2">No data</td><td colspan="2">Blood clot</td><td>A-PRF</td><td>FDBA</td><td>p value</td><td>Blood clot</td><td>A-PRF</td><td colspan="2">FDBA</td><td>p value</td></tr>
<tr><td colspan="2">36.9 ± 14.94</td><td>60.48 ± 9.88</td><td>65.92 ± 10.91</td><td><0.05</td><td colspan="2">9.36 ± 6.49</td><td>10.99 ± 6.39</td><td>9.59 ± 5.38</td><td>>0.05</td><td>53.7 ± 17.79</td><td>28.53 ± 8.66</td><td colspan="2">24.37 ± 9.35</td><td>>0.05</td></tr>
<tr><td rowspan="5">5</td><td rowspan="5">Lavagen et al. [20]</td><td colspan="4">Control (iliac bone graft + PRF)</td><td colspan="3">Test (iliac bone graft + A-PRF)</td><td>p value</td><td colspan="5" rowspan="5">No data</td><td colspan="5" rowspan="5">No data</td></tr>
<tr><td colspan="8">Percentage of newly formed bone</td></tr>
<tr><td colspan="4">51.4 ± 18.4</td><td colspan="3">60.4 ± 10.4</td><td>0.165</td></tr>
<tr><td colspan="8">Mean volume of newly formed bone (cm$^3$)</td></tr>
<tr><td colspan="4">0.2 ± 0.08</td><td colspan="3">0.29 ± 0.09</td><td>0.024</td></tr>
</table>

A-PRF+—advanced platelet-rich fibrin +; DBBM—deproteinized bovine bone mineral; FDBA—freeze-dried bone allograft; and PRGF—plasma rich in growth factors.

### 3.3. A-PRF+'s Effect on Obtained Bone Mineral Density

Four randomized controlled trials that focused on bone density were identified. The incorporation of A-PRF+ into the existing protocols did not yield statistically significant improvements between the test and control groups [13,15–17].

Clark et al. [13] conducted a measurement of bone density using specimens obtained during implant placement surgeries and presented the results in mg/cm$^3$. The addition of A-PRF+ to the FDBA did not result in an improvement in mineral bone density; in fact, it led to a decrease in this parameter (521 $\pm$ 58 mg/cm$^3$ vs. 551 $\pm$ 58 mg/cm$^3$). However, the incorporation of any biomaterial, whether A-PRF+ or FDBA, demonstrated superior outcomes compared to the control group (487 $\pm$ 64 mg/cm$^3$).

Alhaj et al. [15] conducted a voxel count analysis on superimposed X-ray images at the six-month follow-up. The difference between the control group (802.5 voxels) and the test group (827.5 voxels) was not statistically significant. However, the addition of A-PRF+ to the autologous bone graft resulted in an increase in the clinical performance (246.5 with A-PRF+ vs. 167.25 without A-PRF+, mean difference between baseline and 6-month follow-up).

Two of the studies [16,17] employed a comparison of bone density in Hounsfield Units (HUs) at 6-month [16] and 9-month [17] follow-ups. The study by Kalash et al. [17] did not yield statistically significant results at the 9-month control point. The addition of A-PRF+ to the biomaterial (Xenograft) resulted in only a slight improvement in the mean result (496.86 $\pm$ 43.98 HU without A-PRF+ compared with 518.14 $\pm$ 45.24 HU with A-PRF+). Furthermore, Yewale et al. [16] did not present a statistically significant difference, with a total mean difference between the baseline and the 6-month follow-up (1393.1 HU control vs. 1783.1 HU test group). However, the bone density with the addition of A-PRF+ was higher. The results are presented in Table 4. The authors corrected to adhere to the one unit of measurement (HU as most of the X-ray programs enable comparison in these units). Thanks to this, the studies can be compared. But, the limitation is that the grey values, HU and histological mg HA cm$^{-3}$, are not inherently interchangeable. Converting either HU or CBCT grey values to mg cm$^{-3}$ requires a scanner-specific hydroxyapatite phantom calibration, which none of the included studies reported [13,15–17].

**Table 4.** Differences in mineral bone density.

| No | References | Bone Mineral Density | | | | |
|---|---|---|---|---|---|---|
| 1 | Alhaj et al. [15] | | | Autologous bone graft | Autologous bone graft + A-PRF | *p* value |
| | | Day 2 | | 635.25 voxel (−388.72 HU) | 581 voxel (1024 HU) | 0.481 |
| | | 6 months | | 802.5 voxel (−221.5 HU) | 27.5 voxel (996.5 HU) | 0.684 |
| 2 | Clark et al. [13] | A-PRF+ | A-PRF+ and FDBA (1:1 ratio) | FDBA | Blood clot | *p* value |
| | | 493 $\pm$ 70 mg/cm$^3$ (655 HU) | 521 $\pm$ 58 mg/cm$^3$ (701.7 HU) | 551 $\pm$ 58 mg/cm$^3$ (751.7 HU) | 487 $\pm$ 64 mg/cm$^3$ (645 HU) | <0.05 |
| 3 | Kalash et al. [17] | | | Xenograft | Xenograft + A-PRF | *p* value |
| | | Baseline | | 465.71 $\pm$ 48.57 HU | 471.86 $\pm$ 51.64 HU | 0.823 |
| | | 9 months | | 496.86 $\pm$ 43.98 HU | 518.14 $\pm$ 45.24 HU | 0.39 |
| 4 | Yewale et al. [16] | | | Test group | Control group | *p* value |
| | | 6 months | | 1783.1 HU | 1393.1 HU | 0.005 |

A-PRF+—advanced platelet-rich fibrin +; FDBA—freeze-dried bone allograft; and HUs—Hounsfield Units (values in mg/cm$^3$ are shown in brackets for uniformity).

### 3.4. A-PRF+'s Effect on Pocket Depth (PD) and Clinical Attachment Level (CAL)

Two randomized controlled trials considered the alterations in the pocket depths (PDs) and clinical attachment levels (CALs) subsequent to surgical procedures [17,21].

In their study, Ghonima et al. [21] employed biphasic calcium phosphate (BCP), both with and without the addition of A-PRF+. The changes in the PD and CAL were monitored at three-month intervals, up until the ninth month of the final follow-up period. In comparison to the control group, both the BCP and the addition of A-PRF+ demonstrated significantly superior outcomes. At the initial two check-ups, the incorporation of A-PRF+ resulted in a reduction in PD levels, with no notable differences between the BCP and BCP + A-PRF+ groups (1.95 ± 0.9 mm vs. 1.5 ± 0.74 mm at three months, 2.18 ± 1.27 mm vs. 1.95 ± 0.56 mm). At the final follow-up, the PD in the BCP group was observed to be lower than that observed in the A-PRF+ group (2.04 ± 0.96 vs. 2.27 ± 0.71 mm), though the difference in the results between the two groups was not statistically significant. Similarly, the initial CALs remained lower in the group that received A-PRF+ (1.63 ± 0.89 vs. 1.36 ± 1.16 mm), before becoming higher than the sole BCP group at the 6- and 9-month controls (1.68 ± 1.23 vs. 1.77 ± 0.68 mm and 1.68 ± 1.23 vs. 2.13 ± 1.02 mm, respectively).

Kalash et al. [17] only evaluated the PD changes at the 3- and 6-month controls, without assessing the CAL. The addition of A-PRF+ to the protocol resulted in a decrease in the average PD at both the three- and six-month follow-ups (3.39 ± 0.4 vs. 2.96 ± 0.64 mm and 2.89 ± 0.28 vs. 2.57 ± 0.51 mm, respectively). The statistical analysis revealed that the differences between the xenograft and the xenograft + A-PRF+ remained insignificant. The results are presented in Table 5 [17,21].

**Table 5.** Effect of A-PRF+ addition on PD and CAL.

| No | References | Pocket Depth (mm) | | | | Clinical Attachment Level (mm) | | | |
|---|---|---|---|---|---|---|---|---|---|
| | | | BCP | BCP + A-PRF | *p* value | | BCP | BCP + A-PRF | *p* value |
| 1 | Ghonima et al. [21] | 3 months | 1.95 ± 0.9 | 1.5 ± 0.74 | 0.93 | 3 months | 1.63 ± 0.89 | 1.36 ± 1.16 | 0.63 |
| | | 6 months | 2.18 ± 1.27 | 1.95 ± 0.56 | 0.6 | 6 months | 1.68 ± 1.23 | 1.77 ± 0.68 | 0.66 |
| | | 9 months | 2.04 ± 0.96 | 2.27 ± 0.71 | 0.06 | 9 months | 1.68 ± 1.23 | 2.13 ± 1.02 | 0.06 |
| | | | Xenograft | Xenograft + A-PRF | *p* value | | | | |
| 2 | Kalash et al. [17] | 3 months | 3.39 ± 0.4 | 2.96 ± 0.64 | 0.09 | No data | | | |
| | | 6 months | 2.89 ± 0.28 | 2.57 ± 0.51 | 0.173 | | | | |

A-PRF+—advanced platelet-rich fibrin +; BCP—biphasic calcium phosphate.

### 3.5. *A-PRF+'s Effect on Obtained Implant Stability*

Two randomized controlled trials, conducted by Kalash and Angelo, were identified that measured either the primary or secondary implant stability [17,22].

Kalash et al. employed the PerioTest (PTV—PerioTest value; physiological mobility PTV −08 to +09; I grade of mobility +10 to +19; II grade of mobility +20 to +29; III grade of mobility +30 to +50) device to assess the secondary stability at the three- and six-month follow-up periods. The group that utilized A-PRF+ demonstrated superior outcomes in both measurement intervals when compared to the group that solely employed xenograft. At the three-month mark, the xenograft cohort attained a score of −4.14 ± 1.06, while the combination of xenograft and A-PRF+ reached a score of −5.47 ± 1.16. At the final follow-up, six months after surgery, the PerioTest score for the xenograft group exhibited further improvement, reaching −4.51 ± 0.94, while the xenograft + A-PRF+ score demonstrated a similar trend, reaching −6.14 ± 1.27. The differences in the PerioTest score between the groups were statistically significant [17].

Angelo et al. [22] conducted an assessment of the primary stability during the implant insertion procedure. A comparison was made between the native bone and two variants of the EasyGraft product, namely Β-TCP Crystal and Classic. Additionally, the Classic variant was augmented with A-PRF+, thereby facilitating a comparison between the sole EasyGraft Classic. The lowest mean torque values for implant insertion were observed in the native bone (control) group (27.87 ± 6.66 Ncm). All biomaterials demonstrated significantly higher average results, with the highest observed in the EasyGraft Crystal group (52.5 ± 8.15 Ncm). The difference between the EasyGraft Classic (42.51 ± 7.03 Ncm) and Crystal (52.5 ± 8.15 Ncm) variants was found to favor the latter. The incorporation of A-PRF+ into the EasyGraft Classic resulted in an enhanced primary implant stability (42.51 ± 7.03 Ncm vs. 46.89 ± 4.57 Ncm). The results are presented in Table 6.

**Table 6.** Differences in implant stability and insertion torque.

| No | References | Implant Stability and Insertion Torque | | | | | |
|---|---|---|---|---|---|---|---|
| 1 | Kalash et al. [17] | PerioTest (PTV) | Xenograft | | Xenograft + A-PRF+ | | *p* value |
| | | 3 months | −4.14 ± 1.06 | | −5.47 ± 1.16 | | 0.045 |
| | | 6 months | −4.51 ± 0.94 | | −6.14 ± 1.27 | | 0.018 |
| 2 | Angelo et al. [22] | Implant insertion torque (Ncm) | Native bone | Easy Graft Crystal | Easy Graft Classic | Easy Graft Classic + A-PRF | *p* value |
| | | | 27.87 ± 6.66 | 52.5 ± 8.15 | 42.51 ± 7.03 | 46.89 ± 4.57 | <0.05 |

A-PRF+—advanced platelet-rich fibrin plus; PTV—PerioTest value.

## 4. Discussion

The primary objective of this systematic review was to evaluate the effectiveness of A-PRF+'s addition to surgical protocols aimed at managing hard tissues in the oral cavity, both as a sole material and as an additive to grafting materials used for bone augmentation. The present study was conducted based on the premise of randomized clinical trials, which, given their meticulous methodology, yield results that present the highest clinical value. However, it should be noted that such trials can impose certain limitations [23]. One such limitation that was present in our study was the size of the study group of participants, with the lowest patient count being 15 in total. Only two studies presented large patient study groups of over 60 patients. However, it is noteworthy that other studies have amassed sufficient patient counts, thereby mitigating potential biases associated with inadequate participant numbers (328 patients in total across ten studies incorporated in this systematic review). The collective evidence was appraised to be of a moderate to high quality. A notable limitation pertains to the utilization of diverse protocols, each tailored to address the unique requirements of the patient undergoing the respective procedures. Despite the maintenance of the general protocol of A-PRF+ creation, as suggested by Fujioka-Kobayashi et al. [7], there were slight variations in the intraoperative and post-operative methods among the studies. As in daily practice, every surgeon follows a protocol that has been adapted to suit the operator's needs in order to achieve the best possible outcome. This discrepancy complicates the direct comparison between the studies, underscoring the necessity for further research in this area to establish clear guidelines. The post-operative care regimen differed most often during the early healing process, with subsequent follow-ups and clinical/radiological evaluations scheduled at 3 and 6 months.

In the studies presented, A-PRF+ was most frequently utilized as a singular biomaterial in the post-extraction socket, with the objective of preserving its three-dimensional characteristics. The results obtained from the patients' follow-up clearly demonstrated the superiority of the A-PRF+ addition into the protocol. When compared with other blood-derived platelet-rich fibrins, A-PRF+ did not yield any statistically significant results. However, it is noteworthy that A-PRF and A-PRF+ exhibit the highest growth cytokine concentrations among all blood derivatives. Consequently, a comparison with L-PRF can only be made in terms of clinical outcomes in bone management [24]. The increase in the growth cytokines concentration is attributable to the increased neutrophil levels trapped in the fibrin scaffolding during a slower and shorter centrifugal process [5], with A-PRF+ demonstrating superiority over conventional A-PRF. The increased growth cytokines concentration translates to a higher and faster response for the neoangiogenesis and mRNA response, with VEGF highlighted as a main factor for this response [25,26].

In many cases, surgical protocols incorporated the utilization of A-PRF+ in conjunction with bone grafting materials. The rationale for the incorporation of A-PRF+ in bone grafting materials is consistent with the rationale for its sole use, i.e., to maintain the dimensions of the alveolar ridge at the extraction site. The rationale for this choice is based on the density and availability of growth factors provided by the fibrin, which promote the angiogenesis process and mobilize the cellular response to utilize the grafting material as efficiently as possible [27]. The protocol employed by clinicians has been found to influence the grafting material's turnover rate and the subsequent formation of vital and non-vital bone. While the studies have demonstrated a general trend in favor of A-PRF+ utilization, further research is necessary to ascertain its optimal application. In a similar vein to studies solely employing A-PRF+, no statistically significant difference was observed between other blood-derived biomaterials.

The most contentious issue that emerged from the collective analysis of the studies was the assessment of the benefits associated with the collection of patient blood during surgical procedures. This is the one common concern for all of the blood derivatives. In particularly complex cases, heightened scrutiny was applied, given the A-PRF+ preparation's necessity for additional procedures (e.g., blood drawing, centrifugation, and blood clot separation), instruments (e.g., centrifuge machine, glass/plastic tubes, and PRF box), and expenses. As is often the case in clinical practice, the decision regarding the use of A-PRF+ must be made in consultation with the patient, after a thorough discussion of the potential benefits and drawbacks of its implementation in the specific clinical context. As surgical procedures become increasingly intricate, concerns regarding the utilization of A-PRF+ may diminish, as its benefits for bone and grafting material management are indisputable. In such cases, the financial implications, when considered in relation to the overall treatment costs, become a relatively less significant factor.

During the preparation of this systematic review, the primary concern for future studies was the focus on larger study groups and the use of various grafting materials in surgical protocols to better evaluate their effectiveness in oral surgery and implantology. In this study, the bone grafting materials exhibited substantial heterogeneity, as xenografts, allografts, and alloplastic materials were utilized. The paucity of studies conducted on limited study group populations engenders a considerable difficulty in comparing the effects of A-PRF+'s addition to different bone grafts and bone substitute materials. The wide range of clinical protocols created by clinicians to best suit their needs and the various methods of testing the effectiveness of the procedure (CBCT measurements, bone density measurements using grey point or Hounsfield Units, bone sample collections and their immunohistographic evaluation) create another layer of comparative difficulty for the results. For procedures necessitating flap reopening and bone access, the acquisition and subsequent immunohistographic evaluation of bone samples would provide the most bias-proof evidence regarding the clinical outcome, grafting material incorporation and turnover, and bone behavior. For the least invasive approach, it would be valuable to assess the therapeutic success using Hounsfield Units on CBCT and CT and compare their quality, as the procedure of choice to evaluate any three-dimensional changes at the follow-up visit is taking the CBCT or CT.

## 5. Conclusions

In summary, the analyzed studies suggest that A-PRF+ provides a superior stabilization of ridge dimensions following tooth extraction compared to a natural blood clot. Furthermore, when combined with xenografts or alloplastic materials, A-PRF+ may enhance periodontal tissue regeneration, facilitate bone augmentation, and contribute to improved implant stability.

**Author Contributions:** Conceptualization, M.C.; methodology, M.C. and P.A.; formal analysis, M.C. and P.A.; investigation, A.P.; resources, M.C. and P.A.; data curation, M.C. and P.A.; writing—original draft preparation, M.C. and P.A.; writing—review and editing, M.C., P.A. and A.P.; visualization, P.A. and M.C.; supervision, M.C., P.A. and A.P.; project administration, A.P. All authors have read and agreed to the published version of the manuscript.

**Funding:** This research received no external funding.

**Institutional Review Board Statement:** Not required.

**Informed Consent Statement:** Not required.

**Data Availability Statement:** The data presented in this study are available on request from the corresponding author. The data are not publicly available due to privacy restrictions.

**Conflicts of Interest:** The authors declare no conflicts of interest.

## Abbreviations

| | |
|---|---|
| A-PRF | advanced platelet-rich fibrin |
| A-PRF + | advanced platelet-rich fibrin plus |
| BTCP | β-tricalcium phosphate |
| BCP | biphasic calcium phosphate |
| CAL | clinical attachment level |
| DBBM | demineralized bovine bone mineral |
| FDBA | freeze-dried bone allografts |
| HU | Hounsfield Unit |
| PD | pocket depth |
| PRF | platelet-rich fibrin |
| PRGF | plasma rich in growth factors |
| PROSPERO | Prospective Register of Systematic Reviews |

## References

1. Dohan, D.M.; Choukroun, J.; Diss, A.; Dohan, S.L.; Dohan, A.J.J.; Mouhyi, J.; Gogly, B. Platelet-rich fibrin (PRF): A secondgeneration platelet concentrate. Part I: Technological concepts and evolution. *Oral Surg. Oral Med. Oral Pathol. Oral Radiol. Endod.* **2006**, *10*, 37–44. [CrossRef] [PubMed]
2. Brancaccio, Y.; Antonelli, A.; Barone, S.; Bennardo, F.; Fortunato, L.; Giudice, A. Evaluation of local hemostatic efficacy after dental extractions in patients taking antiplatelet drugs: A randomized clinical trial. *Clin. Oral Investig.* **2021**, *25*, 115–1167. [CrossRef]
3. Choukroun, J.; Diss, A.; Simonpieri, A.; Girard, M.O.; Schoeffler, C.; Dohan, S.L.; Dohan, A.J.; Mouhyi, J.; Dohan, D.M. Platelet-rich fibrin (PRF): A second-generation platelet concentrate. Part IV: Clinical effects on tissue healing. *Oral Surg. Oral Med. Oral Pathol. Oral Radiol. Endod.* **2006**, *101*, e56–e60. [CrossRef] [PubMed]
4. Choukroun, J.; Diss, A.; Simonpieri, A.; Girard, M.O.; Schoeffler, C.; Dohan, S.L.; Dohan, A.J.; Mouhyi, J.; Dohan, D.M. Platelet-rich fibrin (PRF): A second-generation platelet concentrate. Part V: Histologic evaluations of PRF effects on bone allograft maturation in sinus lift. *Oral Surg. Oral Med. Oral Pathol. Oral Radiol. Endod.* **2006**, *101*, 299–303. [CrossRef]
5. Ghanaati, S.; Booms, P.; Orlowska, A.; Kubesch, A.; Lorenz, J.; Rutkowski, J.; Landes, C.; Sader, R.; Kirkpatrick, C.; Choukroun, J. Advanced platelet-rich fibrin: A new concept for cell-based tissue engineering by means of inflammatory cells. *J. Oral Implantol.* **2014**, *40*, 679–689. [CrossRef] [PubMed]
6. Choukroun, J. Advanced PRF, and i-PRF: Platalet concentrates or blood concentrates? *J. Periodont. Med. Clin. Pract.* **2014**, *1*, 3.
7. Kobayashi, E.; Flückiger, L.; Fujioka-Kobayashi, M.; Sawada, K.; Sculean, A.; Schaller, B.; Miron, R.J. Comparative Release of Growth Factors from PRP, PRF, and Advanced-PRF. *Clin. Oral Investig.* **2016**, *20*, 2353–2360. [CrossRef]

8. Herrera-Vizcaíno, C.; Dohle, E.; Al-Maawi, S.; Booms, P.; Sader, R.; Kirkpatrick, C.J.; Choukroun, J.; Ghanaati, S. Platelet-Rich Fibrin Secretome Induces Three-Dimensional Angiogenic Activation In Vitro. *Eur. Cell Mater.* **2019**, *37*, 250–264. [CrossRef]
9. Illmilda; Asrianti, D.; Margono, A.; Julianto, I.; Wardoyo, M.P. Advanced Platelet Rich Fibrin (A-PRF) supplemented culture medium for human dental pulp stem cell proliferation. *J. Inter. Dent. Med. Res.* **2019**, *12*, 396–400.
10. Gupta, A.K.; Cole, J.; Deutsch, D.P.; Everts, P.A.; Niedbalski, R.P.; Panchaprateep, R.; Rinaldi, F.; Rose, P.T.; Sinclair, R.; Vogel, J.E.; et al. Platelet-Rich Plasma as a Treatment for Androgenetic Alopecia. *Dermatol. Surg.* **2019**, *45*, 1262–1273. [CrossRef]
11. Everts, P.A.M.; Knape, J.T.; Weibrich, G.; Schönberger, J.P.; Hoffmann, J.; Overdevest, E.P.; Box, H.A.; van Zundert, A. Platelet-rich plasma and platelet gel: A review. *J. Extra Corpor. Technol.* **2006**, *38*, 174–187. [CrossRef] [PubMed]
12. Kargarpour, Z.; Nasirzade, J.; Panahipour, L.; Mitulovi'c, G.; Miron, R.J.; Gruber, R. Platelet-Rich Fibrin Increases BMP2 Expression in Oral Fibroblasts via Activation of TGF-β Signaling. *Int. J. Mol. Sci.* **2021**, *22*, 7935. [CrossRef]
13. Clark, D.; Rajendran, Y.; Paydar, S.; Ho, S.; Cox, D.; Ryder, M.; Dollard, J.; Kao, R.T. Advanced platelet-rich fibrin and freeze-dried bone allograft for ridge preservation: A randomized controlled clinical trial. *J. Periodontol.* **2018**, *89*, 379–387. [CrossRef] [PubMed]
14. Ivanova, V.; Chenchev, I.; Zlatev, S.; Kanazirski, N. Dimensional Ridge Alterations and Histomorphometric Analysis Following Socket Preservation with PRF or Allograft. Randomized Controlled Clinical Study. *J. IMAB* **2019**, *25*, 2853–2861. [CrossRef]
15. Alhaj, F.; Shokry, M.; Attia, N. The efficiency of using advanced platelet rich fibrin–Autogenous bone graft mixture around immediately placed dental implants in mandibular molar region: (Randomized controlled clinical trial). *Egypt. Dent. J.* **2018**, *64*, 2023–2035. [CrossRef]
16. Yewale, M.; Bhat, S.; Kamath, A.; Tamrakar, A.; Patil, V.; Algal, A.S. Advanced platelet-rich fibrin plus and osseous bone graft for socket preservation and ridge augmentation—A randomized control clinical trial. *J. Oral Biol. Craniofac. Res.* **2021**, *11*, 225–233. [CrossRef]
17. Kalash, S.; Aboelsaad, N.; Shokry, M.; Choukroun, J. The efficiency of using advanced PRF-xenograft mixture around immediate implants in the esthetic zone: A randomized controlled clinical trial. *J. Osseointegration* **2017**, *9*, 317–322. [CrossRef]
18. Hartlev, J.; Erik Nørholt, S.; Spin-Neto, R.; Kraft, D.; Schou, S.; Isidor, F. Histology of augmented autogenous bone covered by a platelet-rich fibrin membrane or deproteinized bovine bone mineral and a collagen membrane: A pilot randomized controlled trial. *Clin. Oral Implant. Res.* **2020**, *31*, 694–704. [CrossRef]
19. Dragonas, P.; Prasad, H.S.; Yu, Q.; Mayer, E.T.; Fidel, P.L., Jr. Bone Regeneration in Maxillary Sinus Augmentation Using Advanced Platelet-Rich Fibrin (A-PRF) and Plasma Rich in Growth Factors (PRGF): A Pilot Randomized Controlled Trial. *Int. J. Periodontics Restor. Dent.* **2023**, *43*, 319–327. [CrossRef]
20. Lavagen, N.; Nokovitch, L.; Algrin, A.; Dakpe, S.; Testelin, S.; Devauchelle, B.; Gbaguidi, C. Efficiency of advanced-PRF usage in the treatment of alveolar cleft with iliac bone graft: A retrospective study. *J. Craniomaxillofac. Surg.* **2021**, *49*, 923–928. [CrossRef]
21. Ghonima, J.; El Rashidy, M.; Kotry, G.; Abdelrahman, H. The efficacy of combining advanced platelet-rich fibrin to biphasic alloplast in management of intrabony defects (randomized controlled clinical trial). *Alex. Dent. J.* **2020**, *45*, 8–13. [CrossRef]
22. Angelo, T.; Marcel, W.; Andreas, K.; Izabela, S. Biomechanical Stability of Dental Implants in Augmented Maxillary Sites: Results of a Randomized Clinical Study with Four Different Biomaterials and PRF and a Biological View on Guided Bone Regeneration. *Biomed. Res. Int.* **2015**, *2015*, 850340. [CrossRef]
23. Kostis, J.B.; Dobrzynski, J.M. Limitations of Randomized Clinical Trials. *Am. J. Cardiol.* **2020**, *129*, 109–115. [CrossRef] [PubMed]
24. Caruana, A.; Savina, D.; Macedo, J.P.; Soares, S.C. From Platelet-Rich Plasma to Advanced Platelet-Rich Fibrin: Biological Achievements and Clinical Advances in Modern Surgery. *Eur. J. Dent.* **2019**, *13*, 280–286. [CrossRef] [PubMed]
25. Ferrara, N.; Gerber, H.P. The Role of Vascular Endothelial Growth Factor in Angiogenesis. *Acta Haematol.* **2001**, *106*, 148–156. [CrossRef]
26. Di Alberti, L.; Rossetto, A.; Albanese, M.; D'agostino, A.; De Santis, D.; Bertossi, D.; Nocini, P.F. Expression of Vascular Endothelial Growth Factor (VEGF) mRNA in Healthy Bone Tissue around Implants and in Peri-Implantitis. *Minerva Stomatol.* **2013**, *62*, 1–7. [CrossRef]
27. Sohn, D.S.; Huang, B.; Kim, J.; Park, W.E.; Park, C.C. Utilization of Autologous Concentrated Growth Factors (CGF) Enriched Bone Graft Matrix (Sticky Bone) and CGF-Enriched Fibrin Membrane in Implant Dentistry. *J. Implant. Adv. Clin. Dent.* **2015**, *7*, 11–18.

Journal of
*Functional Biomaterials*

MDPI

*Article*

# Treatment of Medication-Related Osteonecrosis of the Jaw Without and With the Use of Advanced Platelet-Rich Fibrin: A Retrospective Clinical Study

**Paulina Adamska [1,*], Marcin Stasiak [2], Natalia Kobusińska [3], Michał Bartmański [4], Adam Zedler [1] and Michał Studniarek [5]**

1 Division of Oral Surgery, Medical University of Gdańsk, 7 Dębinki Street, 80-211 Gdańsk, Poland; adam.zedler@gumed.edu.pl
2 Division of Orthodontics, Faculty of Medicine, Medical University of Gdańsk, 42c Aleja Zwycięstwa, 80-210 Gdańsk, Poland; marcin.stasiak@gumed.edu.pl
3 University Dental Center, Medical University of Gdańsk, 1a Dębowa Street, 80-204 Gdańsk, Poland; natqus@gumed.edu.pl
4 Institute of Manufacturing and Materials Technology, Faculty of Mechanical Engineering and Ship Technology, Gdańsk University of Technology, 11/12 Gabriela Narutowicza Street, 80-233 Gdańsk, Poland; michal.bartmanski@pg.edu.pl
5 Department of Radiology, Faculty of Medicine, Medical University of Gdańsk, 17 Smoluchowskiego Street, 80-210 Gdańsk, Poland; michal.studniarek@gumed.edu.pl
* Correspondence: paulina.adamska@gumed.edu.pl

**Abstract:** Background: Medication-related osteonecrosis of the jaw (MRONJ) is drug-induced bone destruction that is exposed for a minimum of 6 to 8 weeks in patients who have not received head and neck radiotherapy and who have not been diagnosed with facial bone metastases. MRONJ treatment outcomes are unpredictable. Therefore, alternative treatment methods are being explored, such as blood-derived platelet-rich preparations enriched with growth factors, including advanced platelet-rich fibrin (A-PRF). The presence of growth factors may enhance healing and reduce post-procedure complications. There are no studies examining the effect of A-PRF on the healing of patients with MRONJ. The aim of this study was to retrospectively evaluate treatment outcomes of patients with MRONJ surgically treated without and with the use of A-PRF. Materials and methods: This retrospective study included 28 patients who suffered from osteomyelitis due to MRONJ and underwent surgical treatment between 2019 and 2024. The patients were divided into two groups: the first group received surgical treatment without A-PRF, and the second group received surgical treatment with the application of A-PRF. This study analyzed demographic and clinical data, as well as treatment outcomes. Results: The patients were aged from 43 to 82 years. The most common cause of MRONJ was the administration of zoledronic acid for oncological reasons (22 patients, 78.6%), given intravenously. In 20 patients (71.4%), the antiresorptive treatment lasted longer than three years. The obtained healing distribution was binomial (presence or absence of healing). Estimation of the probability of healing using the maximum likelihood method provided a result of approximately 64%. The probability of ten or more healed patients in the A-PRF group was 41%. A-PRF helps with a probability of 59%, and without A-PRF, it was lower. Concomitantly, the differences between the group with A-PRF and without A-PRF were not statistically significant. Conclusions: The patients with MRONJ should have regular check-ups with radiological examinations at least every six months to detect possible recurrence. Treatment for MRONJ is long and difficult. Treatment of non-advanced lesions, without additional risk factors (such as treatment with zoledronate intravenously for oncological purposes for 3 years), showed a better prognosis. Sometimes, in addition to surgery, it is necessary to consider alternative methods. A-PRF may enhance MRONJ healing. However, there is no evidence of a significant effect of A-PRF on the healing of MRONJ.

**Keywords:** advanced platelet-rich fibrin; A-PRF; autografts; bisphosphonate-associated osteonecrosis of the jaw; bone disease; dentistry; growth factors; osteonecrosis of the jaw

## 1. Introduction

Medication-related osteonecrosis of the jaw (MRONJ) refers to the destruction of jawbones caused by certain drugs. It occurs in patients undergoing or following treatment with antiresorptive or antiangiogenic drugs (e.g., bisphosphonates (BPs; alendronate, etidronate, ibandronate, clodronate, medronate, oxydronate, pamidronate, risedronate, tiludronate, and zoledronate), corticosteroids, denosumab (anti-RANKL), bevacizumab and aflibercept (anti-VEGFR), sunitinib, sorafenib, cabozatinib (tyrosine kinase inhibitors), sirolimus, everolimus (anti-mTOR), and adalimumab (anti-TNF)). In the early stage, avascular necrosis may occur and can only be detected by radiological examination. MRONJ involves the destruction of exposed bone (not covered by oral mucosa), with the exposure persisting for at least 6–8 weeks [1–15]. It develops in patients who have not undergone head and neck radiotherapy and do not have facial bone metastases (Figure 1) [1–3,10–12]. In patients receiving antiresorptive or antiangiogenic drugs, delayed wound healing in the oral cavity and osteonecrosis may occur after surgical procedures or due to other irritating factors, such as sharp edges of teeth or fillings, prosthetic restorations, tooth decay, periodontitis, bone fractures, exostoses, or a pronounced mylohyoid ridge. These complications are most often observed after teeth extractions. Drugs affecting bone turnover are used in the treatment of patients with bone metastases (most frequently for breast or prostate cancer), osteoporosis, fibrous dysplasia, osteogenesis imperfecta, otosclerosis, or multiple myeloma [2,4,7]. The risk of MRONJ is significantly higher in the cases undergoing treatment for tumor metastases when antiresorptive drugs are administered intravenously over 3 years and in those patients receiving additional therapies, such as chemotherapy, steroids, or thalidomide. Additional risk factors are comorbidities (e.g., diabetes, rheumatoid arthritis, calcium deficiency, and hyperparathyroidism) and tobacco smoking [7,10–12].

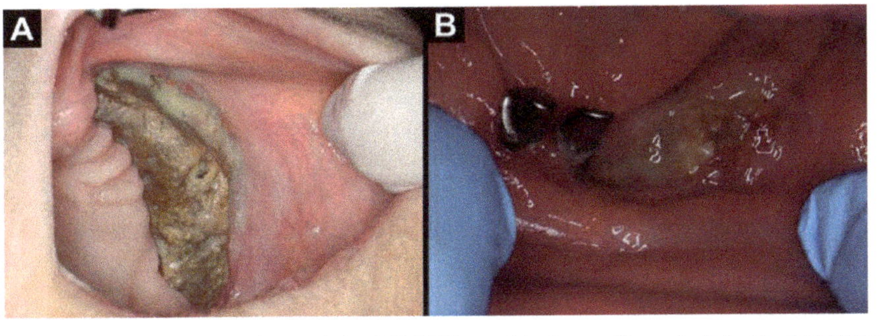


**Figure 1.** Intraoral photographs (**A**,**B**): (**A**) MRONJ located in the maxilla on the left side; (**B**) MRONJ located in the central part of the mandible.

Oral irritants should be eliminated prior to initiating treatment with drugs that affect bone turnover. Dental caries must be treated, dentures adjusted, and damaged teeth removed to prevent MRONJ. Patients should undergo regular clinical and radiological examinations [10–12].

Osteonecrosis occurs approximately in 2/3 of cases in the mandible and in 1/3 in the maxilla. The main symptoms include exposed bone, often accompanied by purulent exudate, pain, soft tissue swelling, an active fistula (intra- or extraoral), halitosis, loosening or loss of teeth, numbness of the lip, a feeling of heaviness, and changes in tactile sensations. In advanced stages, pathological fractures of the mandible may occur. For an extended period, the affected area may remain asymptomatic [2,9,10,16–21].

The presence of osteonecrosis and other MRONJ signs can be assessed in radiological examinations, including dental X-rays, orthopantomography, cone beam computed tomography (CBCT), computed tomography (CT), scintigraphy, single-photon emission computed tomography (SPECT/CT), and positron emission tomography/computed tomography (PET/CT). Non-ionizing diagnostic methods, such as magnetic resonance imaging (MRI), are also utilized. Lesions may appear as radiolucent areas with osteolysis and osteosclerotic defects (commonly referred to as sequestrum with peripheral radiolucency). Unhealed tooth sockets are frequently observed. Pathological cracks, fractures, narrowing of the marrow space, cortical and trabecular bone destruction, and, sometimes, thickening and swelling of the periosteum and soft tissues can be identified on CBCT, CT, or MRI. On T1-weighted MRI, a loss of the fat signal in the bone marrow (typically present in the mandible) may be observed. In contrast, T2-weighted sequences show hypointensity in the bone marrow. Bone scintigraphy shows increased radionuclide uptake at the periphery of the lesion. Similarly, SPECT/CT reveals abnormal radionuclide uptake, such as technetium-99m methylene diphosphonate (99Tcm-MDP) or technetium-99m dicarboxypropane diphosphonate DPD (99Tcm-DPD). However, no uptake is observed in areas of necrosis. In PET/CT imaging, the uptake of fludeoxyglucose F18 (18F-FDG) is higher in inflamed areas compared to necrotic regions. In cases of aseptic necrosis, the affected area cannot be assessed due to the lack of inflammatory processes, which are essential for an accurate analysis [20,22–26].

Histopathological specimens reveal fragments of necrotic bone tissue containing bacteria. Necrotic bone is most commonly infected with *Actinomyces* sp., which are commensal bacteria in the oral cavity. The development of infections and actinomycosis is often associated with a decrease in the host's immunity, commonly observed during oncological treatment. *Actinomycetes* sp. are Gram-positive (+) anaerobes. Bone necrosis creates an anaerobic environment because of vascular degradation and the formation of clots. *Actinomyces* sp. receive an ideal anaerobic environment for their existence. Other bacteria that have been successfully cultured from MRONJ include *Staphylococcus aureus*, *Escherichia coli*, *Streptococcus anginosus*, and *Pseudomonas mendocina* [19,27,28].

Treatment is mainly based on prevention, the elimination of risk factors, and, in advanced stages, on surgical treatment. MRONJ treatment outcomes are unpredictable [1,2,5]. Therefore, alternative treatment methods are being explored, such as blood-derived platelet-rich preparations enriched with growth factors, including advanced platelet-rich fibrin (A-PRF). A-PRF is a blood-derived platelet-rich preparation enriched with growth factors. It belongs to the third generation of preparations derived from blood. A-PRF was developed to obtain advanced blood derivatives rich in growth factors. Compared to classic PRF or L-PRF (leukocyte platelet-rich fibrin), its preparation involves a reduction in the centrifugation speed and a slight increase in the centrifugation time (A-PRF: 1500 rpm/14 min vs. PRF: 3000 rpm/12 min vs. L-PRF: 2700 rpm/12 min). As a result, the biological and physical properties of A-PRF are altered. It contains a greater number of leukocytes and platelets, and the resulting fibrin matrix is more cohesive and exhibits improved adhesion. Additionally, A-PRF is enriched with higher concentrations of cytokines, such as vascular endothelial growth factor (VEGF) and transforming growth factor beta (TGF-β), which enhance healing and stimulate tissue regeneration. An advancement of A-PRF is A-PRF+, which is produced at even lower speeds (1300 rpm/8 min), yielding a preparation with an even higher concentration of growth factors and neutrophils. The presence of growth factors may enhance healing and reduce post-procedure complications, which has been proven in many oral surgery procedures [29–35].

The aim of this study was to retrospectively evaluate the treatment outcomes of patients with medication-related osteonecrosis of the jaw, both without and with the use of advanced platelet-rich fibrin.

## 2. Materials and Methods

The retrospective study included 28 patients (43–82 years old) who suffered from osteomyelitis due to medication-related osteonecrosis of the jaw and underwent surgical treatment at the Division of Oral Surgery, University Dental Center of Medical University of Gdańsk, Poland, between 2019 and 2024. The retrospective research was carried out following the Declaration of Helsinki principles. Approval from the institutional ethics committee was obtained (Independent Bioethics Commission for Research, Medical University of Gdańsk, number of approval KB/510/2024). The patients signed an informed written consent to the surgical procedures.

The analysis considered several factors, including the patients' age, gender, primary disease, duration of antiresorptive or antiangiogenic drug use (less than 3 years or 3 years and above), and route of drug administration (intravenous or oral). Additional factors included the presence of triggers for necrosis development (e.g., tooth extraction, periapical lesion), stage of the disease as classified by the American Association of Oral and Maxillofacial Surgeons (AAOMS, Table 1), and the extent of the lesions in the clinical and radiological imaging (less than 3 cm and 3 cm and above). The analysis also evaluated the type of treatment applied: either surgical removal of the necrotic bone alone or combined with the local administration of growth factors derived from the patient's blood. Treatment success and the duration of the follow-up were assessed. Additionally, several factors influencing a worse prognosis were evaluated, including the use of zoledronic acid, intravenous administration, treatment duration of at least three years, a previous episode of jawbone necrosis, and oncological treatment.

**Table 1.** Classification of MRONJ by the American Association of Oral and Maxillofacial Surgeons.

| | |
|---|---|
| Stage 0 | - Nonexposed bone<br>- Non-specific clinical symptoms (odontalgia, dull pain in maxilla-facial bones/sinuses, altered neurosensory function, teeth loosening, intra- or extraoral swelling)<br>- Changes in radiography (bone resorption not compared with periodontal disease, osteosclerosis) |
| Stage 1 | - Exposed bone or the presence of a purulent fistula<br>- Presence of clinical symptoms<br>- Radiographic changes (bone resorption not attributable to periodontal disease or areas of osteosclerosis) |
| Stage 2 | - Exposed bone or purulent fistula<br>- Presence of clinical symptoms<br>- Changes in radiography (bone resorption not compared with periodontal disease, osteosclerosis) |
| Stage 3 | - Exposed bone or a purulent fistula extending to the alveolar bone<br>- Pathological fractures<br>- Oroantral or oronasal communications or fistulas<br>- Osteolysis extending to the inferior border of the mandible |

Two groups of patients were compared. The first group comprised patients who received surgical treatment involving the removal of necrotic lesions. The second group included patients who, in addition to the removal of necrotic lesions, received advanced platelet-rich fibrin.

### 2.1. Inclusion Criteria

The study included adult patients with MRONJ who had been treated with antiresorptive or antiangiogenic drugs. The eligibility criteria required bone exposure persisting for a minimum of 6–8 weeks, the absence of prior head and neck radiotherapy, and no diagnosis of facial bone metastases. Surgical procedures were performed only on patients in good general condition with normal results in basic diagnostic tests (e.g., blood morphology).

### 2.2. Exclusion Criteria

The study excluded patients with osteonecrosis of the jaw who had not been treated with antiresorptive or antiangiogenic drugs, whose bone exposure persisted for less than 6–8 weeks, who had undergone head and neck radiotherapy, or who were diagnosed with facial bone metastases. Additionally, underage patients and those who did not provide consent for treatment were excluded.

### 2.3. Surgical Procedure

Before the treatment, extraoral and intraoral examinations and radiological diagnostic imaging were performed. Orthopantomography or cone beam computer tomography was assessed using the CS 3D Imaging v3.5.18 software (Carestream Health Inc., Trophy, Croissy-Beaubourg, France). Each patient received 1 g of amoxicillin with cluvuate acid (0.875 g + 0.125 g; Augmentin, GlaxoSmithKline, London, UK) every 12 h one day before the procedure. In patients with allergies, 300 mg of clindamycin was administered every 8 h starting the day before the surgery (Dalacin C, Pfizer, Brooklyn, NY, USA). The antibiotic therapy protocol was implemented according to the Recommendations of the Working Group of the Polish Dental Association and the National Antibiotic Protection Program regarding the use of antibiotics in dentistry [36]. Additionally, the patients rinsed their mouths with a 0.1% chlorhexidine solution (Eludril Classic, Pierre Fabre Oral Care, Lavaur, France).

In the group of patients who received A-PRF before the surgery, 4 tubes of 10 mL venous blood were collected into sterile, anticoagulant-free, glass-coated plastic tubes. After that, the tubes were immediately centrifuged (All Centrifuge, Scilogex, LLC, Rocky Hill, CT, USA). The blood was centrifuged at 1500 rpm for 14 min. Then, the A-PRF clots were put into a PRF box (Quadrostom, Kraków, Poland), and A-PRF plugs were made.

The surgical procedure was performed under local anesthesia administered via injection (Figures 2 and 3). The type of anesthesia depended on the location of the lesion. The nerves were blocked using 4% articaine hydrochloride with 1:100,000 epinephrine (Septodont, Lancaster, PA, USA). A 15c scalpel blade (Swann-Morton, Sheffield, UK) was used, and the mucoperiosteal flap was raised. The epithelialized edges of the mucosa were prepared and debrided. The necrotic bone was curetted, fixed in 10% formalin, and submitted for histopathological examination. The bottom of the lesion was cleaned down to the bleeding bone using a rose head bur (Meisinger, Hager, and Meisinger GmbH, Neuss, Germany) mounted on a surgical straight handpiece (S-11 L G, W&H Dentalwerk Bürmoos GmbH, Bürmoos, Austria) at 40,000 rpm, with abundant irrigation using 0.9% NaCl. The wound was rinsed with 20 mL of 0.5% metronidazole solution (Polpharma SA, Starogard Gdański, Poland). In the A-PRF group, four wound plugs were placed at this stage. The wound was sutured without tension, and hemostasis was achieved. A gauze pad was applied, and the patient was instructed to bite on it and hold it in place for 20 min.

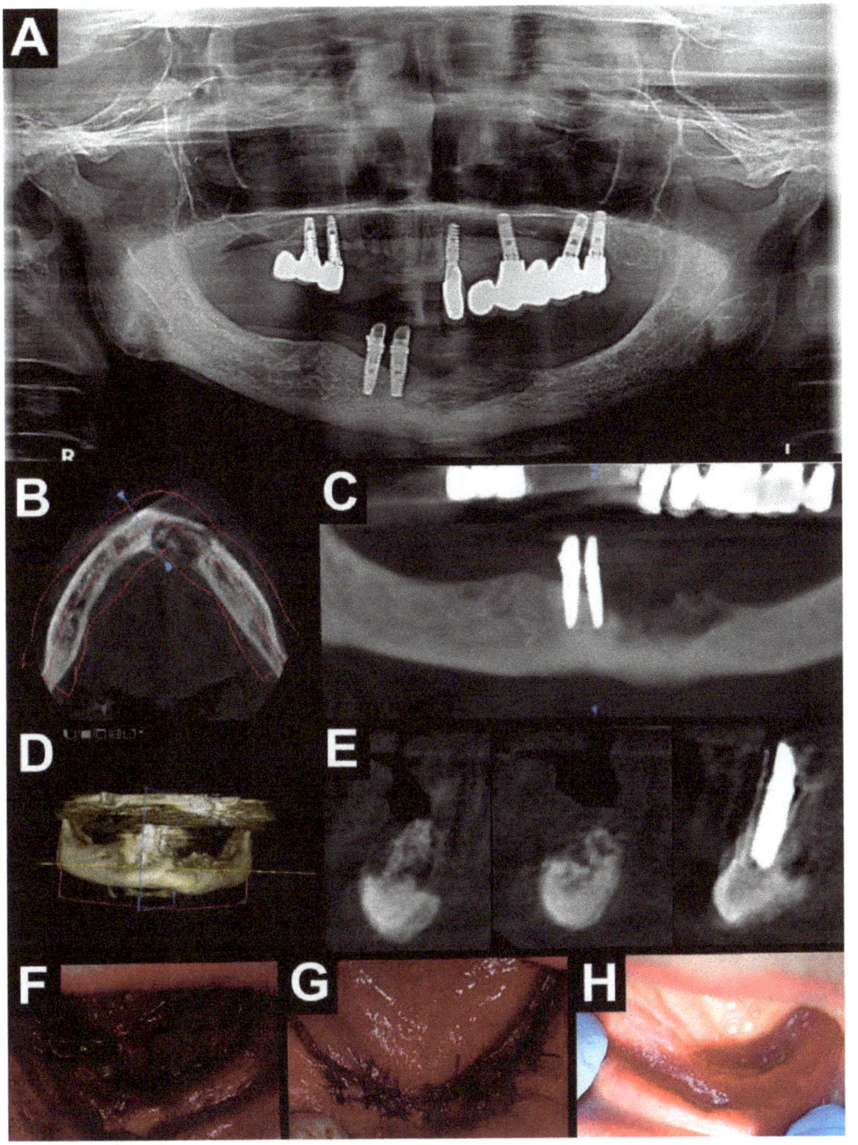


**Figure 2.** Treatment without A-PRF. Radiological diagnostic imaging of MRONJ (**A–E**). (**A**) Orthopantomography—area with osteolysis and osteosclerosis located in the mandible on the anterior area; CBCT (**B–E**). (**B**) Axial view—area with osteolysis and osteosclerosis on the anterior area of the mandible. (**C**) Orthopantomographic reconstruction—area with osteolysis and osteosclerosis on the anterior area of the mandible. (**D**) Pseudo-3D reconstruction—area with osteolysis and osteosclerosis on the anterior area of the mandible. (**E**) Cross-sectional view—area with osteolysis and osteosclerosis on the anterior area of the mandible. Intraoral photographs (**F–H**). (**F**) Visible necrotic bone after mucoperiosteal flap elevation. (**G**) After suturing the wound without tension. (**H**) Healing after 2 weeks (red line—panoramic reconstruction site; yellow line – cross-sectional area line; blue line—cross-sectional area line).

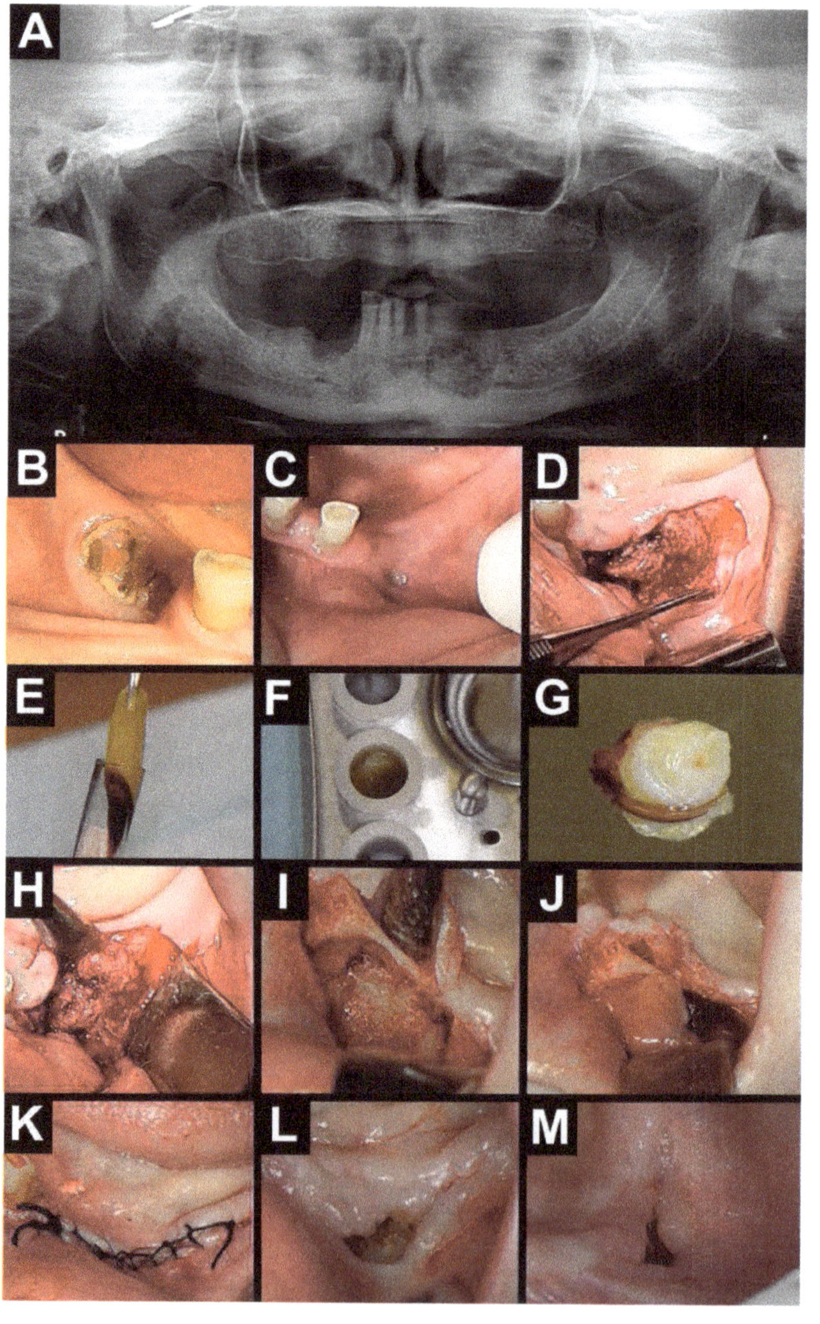


**Figure 3.** Treatment with A-PRF. (**A**) Orthopantomography—area with osteolysis and osteosclerosis on the left and right sides of the mandible. Intraoral photography (**B–D**). (**B**) Exposed necrotic bone on the right side of the mandible. (**C**) Intraoral purulent fistula on the left side of the mandible. (**D**) Visible necrotic bone after mucoperiosteal flap elevation on the left side of the mandible. A-PRF preparation (**E–G**). (**E**) A-PRF clot. (**F**) A-PRF in a PRF box. (**G**) A-PRF plug. Intraoral photographs (**H–M**). (**H,I**) Necrotic tissue was removed, and the bone was bled. (**J**) A-PRF plug placed intra-wound. (**K**) Wound was sutured. (**L**) Healing after 2 weeks. (**M**) Healing after 6 weeks.

Patients continued antibiotic therapy for 14 days in the high-risk group (defined as receiving zoledronic acid, intravenous bisphosphonates, or the drug for at least 3 years, or those with a previous episode of jawbone inflammation or necrosis) and for 7 days in the low-risk group (not meeting these criteria) [36]. The sutures were removed after 14 days.

### *2.4. Post-Operative Evaluation and Follow-Up*

Patients attended regular follow-up appointments: initially at 2 and 6 weeks, then every 3 months, and annually if the lesions healed properly. For non-healing lesions, the frequency of visits was adjusted on an individual basis.

### *2.5. Statistical Analysis*

A statistical analysis was performed using the R software (version 4.4.2, the R Foundation for Statistical Computing Platform, Boston, MA, USA). Patients' age, gender, primary disease, duration of antiresorptive or antiangiogenic drug use (less than 3 years or 3 years and above), route of drug administration (intravenous or oral), the presence of triggers for necrosis development (e.g., tooth extraction, periapical lesion), stage of the disease as classified by the AAOMS, the extent of the lesions in the clinical and radiological imaging (less than 3 cm and 3 cm and above), the type of treatment applied, treatment success, and the duration of follow-ups were quantitatively summarized. Statistical analyses were performed using the maximum likelihood method. The non-parametric ANOVA test was used for a comparative analysis. Statistically significant results were accepted for $p \leq 0.05$.

## 3. Results

### *3.1. Patient Characteristics*

A total of 28 patients with MRONJ were included in this study. At the time of the MRONJ diagnosis, the patients were aged between 43 and 82 years, with a mean age of 70.2. The female group was larger than the male group ($n$ = 21, 75% vs. $n$ = 7, 25%).

Eleven patients (39.3%) had multiple myeloma, seven (25%) had breast cancer, four (14.3%) had prostate cancer, four (14.3%) had osteoporosis, and two (7.1%) had other. All patients were treated exclusively with bisphosphonates—either zoledronic acid or alendronic acid. The most common cause of MRONJ was the administration of zoledronic acid for oncological indications ($n$ = 22, 78.6%). Zoledronic acid was administered intravenously (100% of cases), and alendronic acid was given only orally. In 20 patients (71.4%), the antiresorptive treatment lasted longer than 3 years.

The factors contributing to MRONJ included surgical procedures or infection in the oral cavity. These were tooth extraction ($n$ = 25, 89.3%), inflammatory lesions such as periapical lesions or periodontal disease ($n$ = 2, 7.1%), or the presence of dental implants (1/3.6%).

The location in the mandible was more common than in the maxilla ($n$ = 24, 82.1% vs. $n$ = 5, 17.9%). All patients were diagnosed with stage 3 according to the AAOMS classification. All patients had an intraoral fistula, and two also had an extraoral fistula. In most cases, the lesion size was up to 3 cm ($n$ = 22, 78.6%). Among the oncological patients, lesions up to 3 cm occurred in 100% of the patients in the A-PRF group, and in the group without A-PRF, they were 73%.

Among the non-oncological patients, lesions up to and above 3 cm occurred in the same number of patients in both groups (three patients each). In the A-PRF group, healing was complete in 100% of the patients, and in the non-A-PRF group, it was 66.8%. The average follow-up period was 14.79 months. The demographic and clinical characteristics, divided into groups with and without A-PRF treatment, are presented in Table 2.

**Table 2.** Demographic and clinical features divided into two groups: surgically treated and surgically treated with A-PRF application.

| Features | | Surgical Treatment | Surgical Treatment and Application of A-PRF | $p$ Value |
|---|---|---|---|---|
| Gender | Female | 12 (42.9%) | 9 (32.1%) | 0.2043 |
| | Male | 2 (7.1%) | 5 (17.9%) | |
| Average age at MRONJ diagnosis | Female | 72.3 years | 79.6 years | - |
| | Male | 71 years | 64.1 years | |
| Mean age at MRONJ diagnosis | | 72.07 years | 68.29 years | 0.3928 |
| Primary disease | Oncological disease | 11 (39.3%) | 11 (39.3%) | 0.6369 |
| | Other diseases | 3 (10.7%) | 3 (10.7%) | |
| Type of disease | Multiple myeloma | 3 (10.7%) | 8 (28.6%) | 0.0383 |
| | Breast cancer | 5 (17.9%) | 2 (7.1%) | |
| | Prostate cancer | 3 (10.7%) | 1 (3.6%) | |
| | Osteoporosis | 2 (7.1%) | 2 (7.1%) | |
| | Other | 1 (3.6%) | 1 (3.6%) | |
| Type of drug | Zoledronic acid | 11 (39.3%) | 11 (39.3%) | 0.1708 |
| | Alendronic acid | 3 (10.7%) | 3 (10.7%) | |
| | Other | 0 (0%) | 0 (0%) | |
| Duration of antiresorptive or antiangiogenic drug use | Less than 3 years | 6 (21.4%) | 2 (7.1%) | 0.3097 |
| | 3 years and above | 8 (28.6%) | 12 (42.9%) | |
| Route of administration | Intravenous | 11 (39.3%) | 11 (39.3%) | 0.7993 |
| | Oral | 3 (10.7%) | 3 (10.7%) | |
| Presence of triggers for necrosis development | Yes | 14 (100%) | 14 (100%) | 0.4490 |
| | No | 0 (0%) | 0 (0%) | |
| Localization of MRONJ | Mandible | 12 (42.9%) | 11 (39.3%) | 0.6369 |
| | Maxilla | 2 (7.1%) | 3 (10.7%) | |
| Size of lesion | Less than 3 cm | 10 (35.7%) | 12 (42.9%) | 0.3757 |
| | 3 cm and above | 4 (14.3%) | 2 (7.1%) | |
| Treatment success | Yes | 8 (28.6%) | 10 (35.7%) | 0.4489 |
| | No | 6 (21.4%) | 4 (14.3%) | |
| Mean follow-up period | | 11.1 months | 18.1 months | 0.0521 |

### *3.2. Comparison of the Impact of A-PRF Use in Relation to Demographic and Clinical Features*

The use of A-PRF was compared in relation to demographic and clinical features. No statistical significance was found for the healing and location of lesions (mandible vs. maxilla, $p = 0.4373$) or the size of the lesion (less than 3 cm vs. 3 cm and above, $p = 0.8957$). Statistical significance was obtained only for the presence of healing and the number of additional risk factors (1–5), including zoledronic acid use, intravenous drug administration, drug use for at least three years, a previous episode of jawbone necrosis, or oncological treatment ($p = 0.006989$).

The obtained healing distribution was binomial (presence or absence of healing). Estimation of the probability of healing using the maximum likelihood method provided a result of approximately 64%. The probability of ten or more healed patients in the A-PRF group was 41%. A-PRF helps with a probability of 59%. Concomitantly, the differences between the group with A-PRF and without A-PRF were not statistically significant ($p = 0.449$).

## 4. Discussion

For patients planned to receive oncological or osteoporosis treatment with drugs associated with MRONJ, prophylactic management should be pivotal. Therapy with antiresorptive and antiangiogenic drugs should be started only after eliminating the infection foci in the oral cavity. The patient should undergo a clinical and radiological evaluation or at least an orthopantomographic examination. In complex cases, such as previous endodontic treatments, periodontal disease assessments, or completely impacted teeth, additional tissue evaluation in CBCT may be necessary [10–12].

Due to the challenges in treating medication-related osteonecrosis of the jaw, various therapy approaches are undertaken, including non-surgical and surgical treatments, as well as a combination of both. Alternative techniques are also explored. According to the American Association of Oral and Maxillofacial Surgeons, in the initial stage (stage 0) of the disease, the primary goal is to eliminate irritating factors. Patients are advised to rinse their mouths with 0.12% chlorhexidine. Antibiotic treatments, typically amoxicillin with clavulanic acid or clindamycin, are recommended in the cases of infection. Pain and inflammation management includes analgesics and anti-inflammatory medications. For patients with mucosal damage but no symptoms of infection (stage I), treatment follows the same protocol as in stage 0. In the case of pain, low-level laser therapy (LLLT) can be introduced for biomodulation. In stage II, where bone exposure is accompanied by an active infection, antibacterial, anti-inflammatory, and antiseptic rinses are necessary. At this stage, radiological examination often reveals bone lysis with sequestration. Surgical treatment should be considered as a last resort, as it may exacerbate the lesion progression. Alternatively, a superficial MRONJ debridement can be performed. In the advanced stage (III), surgical treatment is necessary. In stage III, intraoral and extraoral purulent fistulas and, in some cases, pathological fractures are observed. Surgical procedures include sequestrum removal, bone debridement down to healthy bone, or segmental resection [1,2,4,34,35]. In addition to LLLT, other supportive therapies include hyperbaric oxygen therapy (HOT), photobiomodulation, ozone therapy, blood-derived platelet-rich preparations enriched with growth factors, and recombinant human bone morphogenetic proteins (rhBMP). Additional pharmacological treatments involve teriparatide, pentoxifylline, and vitamins E and D. However, alternative methods should not be used as monotherapy but rather as part of a multimodal treatment approach [34,35,37–44].

A-PRF is an autogenous, non-allergenic, and non-immunogenic blood-derived platelet-rich preparation enriched with growth factors. Due to the presence of growth factors, it promotes healing and helps reduce postoperative complications. Wound healing is supported by the gradual release of growth factors (for up to 4 weeks), which enhances regenerative processes, such as cell proliferation and differentiation—particularly of fibroblasts, endothelial cells, and osteoblasts—crucial for both soft and hard tissue regeneration. It also stimulates neovascularization, modulates inflammation, and accelerates the formation of epithelium and connective tissues [29–35].

The available studies about platelet-rich preparations are case reports [45–49] or were used in prophylaxis (e.g., filling the post-extraction socket) [50–55] or other blood products—PRF [46,51,53,56–58] and L-PRF [45,47,48,52,54,59–64]. A-PRF has been used in an animal model [65] and in one case report article [66]. Three original publications analyzed A-PRF+, not A-PRF, as it differs in the production process compared to A-PRF [49,67,68]. In the available studies, the authors reported that the addition of blood-derived agents was a valuable addition, supported healing, and even led to the complete healing of lesions. In a study using PRF, de Almeida Barros Mourão et al. [56] achieved complete treatment success, while in the case of L-PRF, Zelinka et al. [59]—85–94%, Özalp et al. [60]—69%, Aslam et al. [62]—100%, and Yalcin-Ülker [63] achieved a success rate of 80%.

To date, our study is the only one to have investigated A-PRF and its effect on healing in patients with MRONJ. More than 85% of the study participants were oncology patients treated with zoledronic acid administered intravenously. Moreover, the lesions were highly advanced, classified as stage III according to AAOMS. This could have influenced the treatment outcomes. In our study, whether the patient was an oncology patient or not had no impact on the healing process at such advanced stages, regardless of the use of A-PRF. Referring to studies analyzing A-PRF+ (not A-PRF), Giudice et al. [49] reported that 74.5% of their study group consisted of oncology patients, with a predominance of stage II cases according to AAOMS (53%) over stage III (47%). Their findings indicated that A-PRF+ improved the patients' quality of life in the first month after the procedure, reducing both pain and the risk of post-procedure infections. This relationship was not observed in later follow-up periods. Blatt et al. [68] also primarily evaluated oncology patients (84.7%); however, the majority (78.8%) had AAOMS stage I disease, while only 1.9% had stage III disease. Their study found no statistical correlation between healing and the use of A-PRF+. Roman et al. [67] had a heterogeneous study group, including patients with MRONJ who had undergone radiotherapy and cases with a combination of MRONJ and radiotherapy. Among patients with MRONJ, 60.5% were taking oral antiresorptive drugs, which is considered a lower-risk group for MRONJ based on current knowledge. Regarding MRONJ staging, 42.5% of patients were classified as AAOMS stage II, 18.5% as stage I, and 3.7% as stage III. This study concluded that A-PRF+ may serve as an adjunctive method to support wound healing; however, no correlation was found with reductions in disease severity, pain, or oral health-related quality of life.

The early and accurate diagnosis of medication-related osteonecrosis of the jaw is pivotal. Our article broadens the knowledge to include the latest advances in diagnosis and treatment of MRONJ. This article presents diagnostic methods as well as a simple and cheap technique of treating MRONJ using A-PRF. Most importantly, it is non-allergenic and does not cause immunological reactions. Biocompatible healing and the pursuit of nature is called biomimetic treatment. Many patients may require this type of treatment from their doctor due to their beliefs or religion.

The study was limited by the relatively small study group, and the treatment was used only in the advanced stages (which could have resulted in worse treatment results). It would be valuable to compare two groups that are homogeneous in terms of the number of participants and the stage of disease progression but differ in the indication for antiresorptive therapy—oncologic versus non-oncologic patients. A significantly better response to A-PRF treatment was observed in the case of small lesions. Some patients experienced progression of the neoplastic disease, which resulted in a deterioration of their general condition and could have contributed to possible healing failure in the group with or without A-PRF.

Future research directions should focus on large cohort studies. It is important to remember to support the treatment with antibiotic therapy. An interesting solution would be a combined treatment with pentoxifylline and tocopherol, then performing a surgical procedure—bone debridement with additional bone ablation with a laser and the application of growth factors in an antibiotic cover, followed by laser biostimulation.

## 5. Conclusions

The patients with medication-related osteonecrosis of the jaw should undergo regular check-ups, including radiological examinations, at least every six months to detect possible

recurrence. MRONJ treatment is prolonged and challenging. Non-advanced lesions, in the absence of additional risk factors (such as zoledronate treatment administered intravenously for oncological reasons over a period of more than 3 years), tend to have a better prognosis. In some cases, alternative therapies may support surgical treatments. A-PRF may enhance MRONJ healing. However, there is no evidence of a significant effect of A-PRF on the healing of MRONJ. We believe that increasing awareness and understanding of these therapeutic approaches is essential. Further studies on the use of A-PRF in the treatment of MRONJ are needed in a homogeneous, large study group in oncological and non-oncological patients.

**Author Contributions:** Conceptualization, P.A.; methodology, P.A.; software, P.A. and M.S. (Marcin Stasiak); validation, P.A. and M.S. (Marcin Stasiak); formal analysis, P.A.; investigation, P.A.; resources, P.A. and N.K.; data curation, P.A. and N.K.; writing—original draft preparation, P.A.; writing—review and editing, P.A., M.S. (Marcin Stasiak), M.B. and M.S. (Michał Studniarek); visualization, P.A.; supervision, P.A., A.Z. and M.S. (Michał Studniarek); project administration, P.A. and M.S. (Michał Studniarek). All authors have read and agreed to the published version of the manuscript.

**Funding:** This research received no external funding.

**Institutional Review Board Statement:** This study was conducted according to the guidelines of the Declaration of Helsinki and was approved by the Independent Bioethics Committee for Scientific Research at the Medical University of Gdańsk, Poland (protocol code KB/510/2024 and date of approval 12 December 2024).

**Informed Consent Statement:** Informed consent was obtained from all subjects involved in the study.

**Data Availability Statement:** The original contributions presented in the study are included in the article, further inquiries can be directed to the corresponding author.

**Acknowledgments:** Paulina Adamska and Natalia Kobusińska would like to thank Antoni Jusyk for his guidance and support in helping us develop the art of surgery as our mentor.

**Conflicts of Interest:** The authors declare no conflicts of interest.

## Abbreviations

| | |
|---|---|
| 99Tcm-DPD | technetium-99m dicarboxypropane diphosphonate |
| 99Tcm-MDP | technetium-99m methylene diphosphonate |
| A-PRF | advanced platelet-rich fibrin |
| A-PRF+ | advanced platelet-rich fibrin + |
| AAOMS | American Association of Oral and Maxillofacial Surgeons |
| anti-mTOR | anti-mammalian target of rapamycin |
| anti-RANKL | anti-receptor activator of nuclear factor-kappaB ligand antibody |
| anti-VEGFR | anti-vascular-endothelial growth factor receptor |
| BPs | bisphosphonates |
| CBCT | cone beam computed tomography |
| F18-FDG | fludeoxyglucose F18 |
| HOT | hyperbaric oxygen therapy |
| L-PRF | leukocyte platelet-rich fibrin |
| LLLT | low-level laser therapy |
| MRI | magnetic resonance imaging |
| MRONJ | medication-related osteonecrosis of the jaw |
| PET/CT | positron emission tomography/computed tomography |
| PRF | platelet-rich fibrin |
| rhBMP | recombinant human bone morphogenetic proteins |
| SPECT/CT | single-photon emission computed tomography |

## References


1. Marx, R.E. *Oral and Intravenous Bisphosphonate-Induced Osteonecrosis of the Jaws. History, Etiology, Prevention, and Treatment*; Quintessence Publishing Co., Inc.: Batavia, NY, USA, 2007; pp. 9–96.
2. American Dental Association Council on Scientific Affairs. Dental management of patients receiving oral bisphosphonatetherapy: Expert panel recommendations. *J. Am. Dent. Assoc.* **2006**, *137*, 1144–1150. [CrossRef]
3. Sarin, J.; DeRossi, S.S.; Akintoye, S.O. Updates on bisphosphonates and potential pathobiology of bisphosphonate-induced jaw osteonecrosis. *Oral Dis.* **2008**, *14*, 277–285. [CrossRef] [PubMed]
4. Kim, D.W.; Jung, Y.S.; Park, H.S.; Jung, H.D. Osteonecrosis of the jaw related to everolimus: A case report. *Br. J. Oral Maxillofac. Surg.* **2013**, *51*, e302–e304. [CrossRef] [PubMed]
5. Ruggiero, S.L.; Dodson, T.B.; Fantasia, J.; Goodday, R.; Aghaloo, T.; Mehrotra, B.; O'Ryan, F.; American Association of Oral and Maxillofacial Surgeons. American Association of Oral and Maxillofacial Surgeons position paper on medication-related osteonecrosis of the jaw-2014 update. *J. Oral Maxillofac. Surg.* **2014**, *72*, 1938–1956. [CrossRef] [PubMed]
6. Katsarelis, H.; Shah, N.P.; Dhariwal, D.K.; Pazianas, M. Infection and medication-related osteonecrosis of the jaw. *J. Dent. Res.* **2015**, *94*, 534–539. [CrossRef]
7. Rosella, D.; Papi, P.; Giardino, R.; Cicalini, E.; Piccoli, L.; Pompa, G. Medication-related osteonecrosis of the jaw: Clinical and practical guidelines. *J. Int. Soc. Prev. Community Dent.* **2016**, *6*, 97–104. [CrossRef]
8. Abed, H.H.; Al-Sahafi, E.N. The role of dental care providers in the management of patients prescribed bisphosphonates: Brief clinical guidance. *Gen. Dent.* **2018**, *66*, 18–24.
9. Nancollas, G.H.; Tang, R.; Phipps, R.J.; Henneman, Z.; Gulde, S.; Wu, W.; Mangood, A.; Russell, R.G.; Ebetino, F.H. Novel insights into actions of bisphosphonates on bone: Differences in interactions with hydroxyapatite. *Bone* **2006**, *38*, 617–627. [CrossRef]
10. Campisi, G.; Mauceri, R.; Bertoldo, F.; Bettini, G.; Biasotto, M.; Colella, G.; Consolo, U.; Di Fede, O.; Favia, G.; Fusco, V.; et al. Medication-Related Osteonecrosis of Jaws (MRONJ) Prevention and Diagnosis: Italian Consensus Update 2020. *Int. J. Environ. Res. Public Health* **2020**, *17*, 5998. [CrossRef]
11. Fusco, V.; Campisi, G.; Carcieri, P.; Fagioli, F.; Bertetto, O.; Mignogna, M.D.; Bedogni, A. Osteonecrosis of Jaw Related to Bisphosphonates and Other Drugs—Prevention, Diagnosis, Pharmacovigilance, Treatment: A 2021 Web Event. *Oral* **2022**, *2*, 137–147. [CrossRef]
12. De Cicco, D.; Boschetti, C.E.; Santagata, M.; Colella, G.; Staglianò, S.; Gaggl, A.; Bottini, G.B.; Vitagliano, R.; D'amato, S. Medication-Related Osteonecrosis of the Jaws: A Comparison of SICMF–SIPMO and AAOMS Guidelines. *Diagnostics* **2023**, *13*, 2137. [CrossRef]
13. Bovari-Biri, J.; Miskei, J.A.; Kover, Z.; Steinerbrunner-Nagy, A.; Kardos, K.; Maroti, P.; Pongracz, J.E. Advancements in Bone Replacement Techniques–Potential Uses After Maxillary and Mandibular Resections Due to Medication-Related Osteonecrosis of the Jaw (MRONJ). *Cells* **2025**, *14*, 145. [CrossRef] [PubMed]
14. Laputková, G.; Talian, I.; Schwartzová, V. Medication-Related Osteonecrosis of the Jaw: A Systematic Review and a Bioinformatic Analysis. *Int. J. Mol. Sci.* **2023**, *24*, 16745. [CrossRef]
15. Caggiano, M.; Di Spirito, F.; Acerra, A.; Galdi, M.; Sisalli, L. Multiple-Drugs-Related Osteonecrosis of the Jaw in a Patient Affected by Multiple Myeloma: A Case Report. *Dent. J.* **2023**, *11*, 104. [CrossRef]
16. Kalra, S.; Jain, V. Dental complications and management of patients on bisphosphonate therapy: A review article. *J. Oral Biol. Craniofac. Res.* **2012**, *3*, 25–30. [CrossRef] [PubMed]
17. Sarasquete, M.E.; González, M.; San Migel, J.F.; García-Sanz, R. Bisphosphonate-related osteonecrosis: Genetic and acquired risk factors. *Oral Dis.* **2009**, *15*, 382–387. [CrossRef] [PubMed]
18. Nicolatou-Galitis, O.; Schiødt, M.; Mendes, R.A.; Ripamonti, C.; Hope, S.; Drudge-Coates, L.; Niepel, D.; Van den Wyngaert, T. Medication-related osteonecrosis of the jaw: Definition and best practice for prevention, diagnosis, and treatment. *Oral Surg. Oral Med. Oral Pathol. Oral Radiol.* **2019**, *127*, 117–135. [CrossRef]
19. Lam, D.K.; Sándor, G.K.; Holmes, H.I.; Evans, A.W.; Clokie, C.M. A review of bisphosphonate-associated osteonecrosis of the jaws and its management. *J. Can. Dent. Assoc.* **2007**, *73*, 417–422.
20. Ruggiero, S.L. Guidelines for the diagnosis of bisphosphonate-related osteonecrosis of the jaw (BRONJ). *Clin. Cases Miner. Bone Metab.* **2012**, *4*, 37–42.
21. Gutta, R.; Louis, P.J. Bisphosphonates and osteonecrosis of the jaws: Science and rationale. *Oral Surg. Oral Med. Oral Pathol. Oral Radiol. Endod.* **2007**, *104*, 186–193. [CrossRef]
22. Haworth, A.E.; Webb, J. Skeletal complications of bisphosphonate use: What the radiologist should know. *Br. J. Radiol.* **2012**, *85*, 1333–1342. [CrossRef]
23. Andreu-Arasa, V.C.; Chapman, M.N.; Kuno, H.; Fujita, A.; Sakai, O. Craniofacial manifestations of systemic disorders: CT and MR imaging findings and imaging approach. *Radiographics* **2018**, *38*, 890–911. [CrossRef] [PubMed]
24. Ko, Y.Y.; Yang, W.-F.; Leung, Y.Y. The Role of Cone Beam Computed Tomography (CBCT) in the Diagnosis and Clinical Management of Medication-Related Osteonecrosis of the Jaw (MRONJ). *Diagnostics* **2024**, *14*, 1700. [CrossRef] [PubMed]

25. Berg, B.-I.; Mueller, A.A.; Augello, M.; Berg, S.; Jaquiéry, C. Imaging in Patients with Bisphosphonate-Associated Osteonecrosis of the Jaws (MRONJ). *Dent. J.* **2016**, *4*, 29. [CrossRef]
26. Kojima, Y.; Soutome, S.; Otsuru, M.; Hayashida, S.; Sakamoto, Y.; Sawada, S.; Umeda, M. Factors Exacerbating Clinical Symptoms and CT Findings in Patients with Medication-Related Osteonecrosis of the Jaw Receiving Conservative Therapy: A Multicenter Retrospective Study of 53 Cases. *Int. J. Environ. Res. Public Health* **2022**, *19*, 7854. [CrossRef]
27. Ibrahim, N.; Apandi, N.I.M.; Shuhardi, S.A.; Ramli, R. *Actinomyces* sp. Presence in the Bone Specimens of Patients with Osteonecrosis of the Jaw: The Histopathological Analysis and Clinical Implication. *Antibiotics* **2022**, *11*, 1067. [CrossRef] [PubMed]
28. Ciobanu, G.A.; Mogoantă, L.; Camen, A.; Ionescu, M.; Vlad, D.; Staicu, I.E.; Munteanu, C.M.; Gheorghiță, M.I.; Mercuț, R.; Sin, E.C.; et al. Clinical and Histopathological Aspects of MRONJ in Cancer Patients. *J. Clin. Med.* **2023**, *12*, 3383. [CrossRef]
29. Chmielewski, M.; Pilloni, A.; Adamska, P. Application of Advanced Platelet-Rich Fibrin in Oral and Maxillo-Facial Surgery: A Systematic Review. *J. Funct. Biomater.* **2024**, *15*, 377. [CrossRef]
30. Adamska, P.; Kaczoruk-Wieremczuk, M.; Pylińska-Dąbrowska, D.; Stasiak, M.; Bartmański, M.; Zedler, A.; Studniarek, M. Treatment of Oroantral Communication and Fistulas with the Use of Blood-Derived Platelet-Rich Preparations Rich in Growth Factors: A Systematic Review. *Int. J. Mol. Sci.* **2024**, *25*, 11507. [CrossRef]
31. Adamska, P.; Pylińska-Dąbrowska, D.; Stasiak, M.; Kaczoruk-Wieremczuk, M.; Kozłowska, E.; Zedler, A.; Studniarek, M. Treatment of Odontogenic Maxillary Sinusitis with the Use of Growth Factors in Advanced Platelet-Rich Fibrin for Immediate Closure of Oro-Antral Communication: A Case Report. *Int. J. Mol. Sci.* **2024**, *25*, 4339. [CrossRef]
32. Chmielewski, M.; Pilloni, A.; Adamska, P. Advanced Platelet-Rich Fibrin Plus (A-PRF+) as an Additive to Hard Tissue Managing Protocols in Oral Surgery: A Systematic Review. *J. Funct. Biomater.* **2025**, *16*, 145. [CrossRef] [PubMed]
33. Adamska, P.; Pylińska-Dąbrowska, D.; Stasiak, M.; Sobczak-Zagalska, H.; Jusyk, A.; Zedler, A.; Studniarek, M. Tooth Autotransplantation, Autogenous Dentin Graft, and Growth Factors Application: A Method for Preserving the Alveolar Ridge in Cases of Severe Infraocclusion—A Case Report and Literature Review. *J. Clin. Med.* **2024**, *13*, 3902. [CrossRef]
34. Lopez-Jornet, P.; Perez, A.S.; Arturo, S.P.; Rui, A.M.; Aurelio, T. Medication-related osteonecrosis of the jaw: Is autologous platelet concentrate application effective for prevention and treatment? A systematic review. *J. Craniomaxillofac. Surg.* **2016**, *44*, 1067–1072. [CrossRef]
35. Frutuoso, F.; Freitas, F.; Vilares, M.; Francisco, H.; Marques, D.; Caramês, J.; Moreira, A. Medication-Related Osteonecrosis of the Jaw: A Systematic Review of Case Reports and Case Series. *Diseases* **2024**, *12*, 205. [CrossRef] [PubMed]
36. Kaczmarczyk, T.; Babiuch, K.; Bołtacz-Rzepkowska, E.; Dominiak, M.; Konopka, T.; Lipski, M.; Olczak-Kowalczyk, D.; Szeląg, A.; Szuta, M.; Hryniewicz, W. *Rekomendacje Grupy Roboczej Polskiego Towarzystwa Stomatologicznego i Narodowego Programu Ochrony Antybiotyków w Zakresie Stosowania Antybiotyków w Stomatologii*; Narodowy Instytut Leków: Warszawa, Poland, 2019.
37. Momesso, G.A.C.; Lemos, C.A.A.; Santiago-Júnior, J.F.; Faverani, L.P.; Pellizzer, E.P. Laser surgery in management of medication-related osteonecrosis of the jaws: A meta-analysis. *Oral Maxillofac. Surg.* **2020**, *24*, 133–144. [CrossRef]
38. Freiberger, J.J. Utility of hyperbaric oxygen in treatment of bisphosphonate-related osteonecrosis of the jaws. *Oral Maxillofac. Surg.* **2009**, *67*, 96–106. [CrossRef] [PubMed]
39. Nica, D.F.; Riviș, M.; Roi, C.I.; Todea, C.D.; Duma, V.-F.; Sinescu, C. Complementarity of Photo-Biomodulation, Surgical Treatment, and Antibiotherapy for Medication-Related Osteonecrosis of the Jaws (MRONJ). *Medicina* **2021**, *57*, 145. [CrossRef]
40. Ripamonti, C.I.; Cislaghi, E.; Mariani, L.; Maniezzo, M. Efficacy and safety of medical ozone (O(3)) delivered in oil suspension applications for the treatment of osteonecrosis of the jaw in patients with bone metastases treated with bisphosphonates: Preliminary results of a phase I-II study. *Oral Oncol.* **2011**, *47*, 185–190. [CrossRef]
41. Kwon, Y.-D.; Kim, D.-Y. Role of Teriparatide in Medication-Related Osteonecrosis of the Jaws (MRONJ). *Dent. J.* **2016**, *4*, 41. [CrossRef]
42. Jung, J.; Yoo, H.Y.; Kim, G.T.; Lee, J.W.; Lee, Y.A.; Kim, D.Y.; Kwon, Y.D. Short-term teriparatide and recombinant human bone morphogenetic protein-2 for regenerative approach to medication-related osteonecrosis of the jaw: A preliminary study. *J. Bone Miner. Res.* **2017**, *32*, 2445–2452. [CrossRef]
43. Colapinto, G.; Goker, F.; Nocini, R.; Albanese, M.; Nocini, P.F.; Sembronio, S.; Argenta, F.; Robiony, M.; Del Fabbro, M. Outcomes of a Pharmacological Protocol with Pentoxifylline and Tocopherol for the Management of Medication-Related Osteonecrosis of the Jaws (MRONJ): A Randomized Study on 202 Osteoporosis Patients. *J. Clin. Med.* **2023**, *12*, 4662. [CrossRef] [PubMed]
44. Michalak, F.; Dominiak, M.; Kiryk, J.; Popecki, P.; Kubicki, D.; Matys, J.; Grzech-Leśniak, K. The Influence of Vitamin D Levels and Supplementation on the Treatment of Patients Affected by MRONJ. *Appl. Sci.* **2025**, *15*, 670. [CrossRef]
45. Pardo-Zamora, G.; Martínez, Y.; Moreno, J.A.; Ortiz-Ruíz, A.J. Treatment of Stage 2 Medication-Induced Osteonecrosis of the Jaw: A Case Series. *Int. J. Environ. Res. Public Health* **2021**, *18*, 1018. [CrossRef]
46. Bennardo, F.; Bennardo, L.; Del Duca, E.; Patruno, C.; Fortunato, L.; Giudice, A.; Nisticò, S.P. Autologous platelet-rich fibrin injections in the management of facial cutaneous sinus tracts secondary to medication-related osteonecrosis of the jaw. *Dermatol. Ther.* **2020**, *33*, e13334. [CrossRef] [PubMed]

47. Hao, L.; Tian, Z.; Li, S.; Yan, K.; Xue, Y. Osteonecrosis of the jaw induced by bisphosphonates therapy in bone metastases patient: Case report and literature review. *Oral Oncol.* **2022**, *128*, 105852. [CrossRef] [PubMed]
48. Bouland, C.; Meuleman, N.; Widelec, J.; Keiani-Mothlagh, K.; Voisin, C.; Lagneaux, L.; Philippart, P. Case reports of medication-related osteonecrosis of the jaw (MRONJ) treated with uncultured stromal vascular fraction and L-PRF. *J. Stomatol. Oral Maxillofac. Surg.* **2021**, *122*, 212–218. [CrossRef]
49. Giudice, A.; Antonelli, A.; Muraca, D.; Fortunato, L. Usefulness of advanced-platelet rich fibrin (A-PRF) and injectable-platelet rich fibrin (i-PRF) in the management of a massive medication-related osteonecrosis of the jaw (MRONJ): A 5-years follow-up case report. *Indian J. Dent. Res.* **2020**, *31*, 813–818. [CrossRef]
50. Asaka, T.; Ohga, N.; Yamazaki, Y.; Sato, J.; Satoh, C.; Kitagawa, Y. Platelet-rich fibrin may reduce the risk of delayed recovery in tooth-extracted patients undergoing oral bisphosphonate therapy: A trial study. *Clin. Oral Investig.* **2017**, *21*, 2165–2172. [CrossRef]
51. Miranda, M.; Gianfreda, F.; Raffone, C.; Antonacci, D.; Pistilli, V.; Bollero, P. The Role of Platelet-Rich Fibrin (PRF) in the Prevention of Medication-Related Osteonecrosis of the Jaw (MRONJ). *Biomed. Res. Int.* **2021**, *2021*, 4948139. [CrossRef]
52. Besi, E.; Pitros, P. The role of leukocyte and platelet-rich fibrin in the prevention of medication-related osteonecrosis of the jaw, in patients requiring dental extractions: An observational study. *Oral Maxillofac. Surg.* **2024**, *28*, 785–793. [CrossRef]
53. Alrmali, A.; Saleh, M.H.A.; Kurdi, S.M.S.; Sabri, H.; Meghil, M.M.; Wang, H.L. Prevention and management of drug-induced osteonecrosis of the jaws using platelet-rich fibrin: A clinical feasibility study. *Clin. Exp. Dent. Res.* **2023**, *9*, 791–798. [CrossRef] [PubMed]
54. Parise, G.K.; Costa, B.N.; Nogueira, M.L.; Sassi, L.M.; Schussel, J.L. Efficacy of fibrin-rich platelets and leukocytes (L-PRF) in tissue repair in surgical oral procedures in patients using zoledronic acid-case-control study. *Oral Maxillofac. Surg.* **2023**, *27*, 507–512. [CrossRef] [PubMed]
55. Şahin, O.; Tatar, B.; Ekmekcioğlu, C.; Aliyev, T.; Odabaşı, O. Prevention of medication related osteonecrosis of the jaw after dentoalveolar surgery: An institution's experience. *J. Clin. Exp. Dent.* **2020**, *12*, e771–e776. [CrossRef] [PubMed]
56. de Almeida Barros Mourão, C.F.; Calasans-Maia, M.D.; Del Fabbro, M.; Le Drapper Vieira, F.; Coutinho de Mello Machado, R.; Capella, R.; Miron, R.J.; Gomes Alves, G. The use of Platelet-rich Fibrin in the management of medication-related osteonecrosis of the jaw: A case series. *J. Stomatol. Oral Maxillofac. Surg.* **2020**, *121*, 84–89. [CrossRef]
57. Giudice, A.; Barone, S.; Giudice, C.; Bennardo, F.; Fortunato, L. Can platelet-rich fibrin improve healing after surgical treatment of medication-related osteonecrosis of the jaw? A pilot study. *Oral Surg. Oral Med. Oral Pathol. Oral Radiol.* **2018**, *126*, 390–403. [CrossRef]
58. Inchingolo, F.; Cantore, S.; Dipalma, G.; Georgakopoulos, I.; Almasri, M.; Gheno, E.; Motta, A.; Marrelli, M.; Farronato, D.; Ballini, A.; et al. Platelet rich fibrin in the management of medication-related osteonecrosis of the jaw: A clinical and histopathological evaluation. *J. Biol. Regul. Homeost. Agents* **2017**, *31*, 811–816.
59. Zelinka, J.; Blahak, J.; Perina, V.; Pacasova, R.; Treglerova, J.; Bulik, O. The use of platelet-rich fibrin in the surgical treatment of medication-related osteonecrosis of the jaw: 40 patients prospective study. *Biomed. Pap. Med. Fac. Univ. Palacky. Olomouc. Czech Repub.* **2021**, *165*, 322–327. [CrossRef]
60. Özalp, Ö.; Yıldırımyan, N.; Öztürk, C.; Kocabalkan, B.; Şimşek Kaya, G.; Sindel, A.; Altay, M.A. Promising results of surgical management of advanced medication related osteonecrosis of the jaws using adjunctive leukocyte and platelet rich fibrin. *BMC Oral Health* **2021**, *21*, 613. [CrossRef]
61. Park, J.H.; Kim, J.W.; Kim, S.J. Does the Addition of Bone Morphogenetic Protein 2 to Platelet-Rich Fibrin Improve Healing After Treatment for Medication-Related Osteonecrosis of the Jaw? *J. Oral Maxillofac. Surg.* **2017**, *75*, 1176–1184. [CrossRef]
62. Aslam, R.D.; Pitros, P.; Liew, J.; Besi, E. The adjunctive use of Leukocyte-Platelet Rich Fibrin (L-PRF) in the management of Medication Related Osteonecrosis of the Jaw (MRONJ): A retrospective observational study. *Oral Maxillofac. Surg.* **2024**, *28*, 1605–1615. [CrossRef]
63. Yalcin-Ülker, G.M.; Duygu, G.; Tanan, G.; Cakir, M.; Meral, D.G. Use of Leukocyte-rich and Platelet-rich Fibrin (L-PRF) Adjunct to Surgical Debridement in the Treatment of Stage 2 and 3 Medication-Related Osteonecrosis of the Jaw. *J. Craniofac. Surg.* **2023**, *34*, 1039–1044. [CrossRef] [PubMed]
64. Tenore, G.; Zimbalatti, A.; Rocchetti, F.; Graniero, F.; Gaglioti, D.; Mohsen, A.; Caputo, M.; Lollobrigida, M.; Lamazza, L.; De Biase, A.; et al. Management of Medication-Related Osteonecrosis of the Jaw (MRONJ) Using Leukocyte- and Platelet-Rich Fibrin (L-PRF) and Photobiomodulation: A Retrospective Study. *J. Clin. Med.* **2020**, *9*, 3505. [CrossRef]
65. Jamalpour, M.R.; Shahabi, S.; Baghestani, M.; Shokri, A.; Jamshidi, S.; Khazaei, S. Complementarity of surgical therapy, photobiomodulation, A-PRF and L-PRF for management of medication-related osteonecrosis of the jaw (MRONJ): An animal study. *BMC Oral Health* **2022**, *22*, 241. [CrossRef] [PubMed]
66. Asfour, M.A.R.; Aljoujou, A.A.; Saifo, M.S.; Jabban, H.A.L. The use of advanced-platelet rich fibrin (A-PRF) in the management of medication-related osteonecrosis of the jaw (MRONJ): A case report. *Clin. Case Rep.* **2023**, *11*, e8259. [CrossRef] [PubMed]

67. Roman, R.C.; Moldovan, M.A.; Pop, L.S.; Megieșan, S.; Faur, C.I. Platelet-Rich Fibrin Treatment Evaluation in Patients with Medication-Related Osteonecrosis of the Jaw and Osteoradionecrosis. *J. Clin. Med.* **2024**, *13*, 3473. [CrossRef]
68. Blatt, S.; Krüger, M.; Kämmerer, P.W.; Thiem, D.G.E.; Matheis, P.; Eisenbeiß, A.-K.; Wiltfang, J.; Al-Nawas, B.; Naujokat, H. Non-Interventional Prospective Observational Study of Platelet Rich Fibrin as a Therapy Adjunctive in Patients with Medication-Related Osteonecrosis of the Jaw. *J. Clin. Med.* **2022**, *11*, 682. [CrossRef]

MDPI AG
Grosspeteranlage 5
4052 Basel
Switzerland
Tel.: +41 61 683 77 34

*Journal of Functional Biomaterials* Editorial Office
E-mail: jfb@mdpi.com
www.mdpi.com/journal/jfb

www.ingramcontent.com/pod-product-compliance
Lightning Source LLC
Chambersburg PA
CBHW061629210926
51850CB00009B/2485
*9783725852796*